The
Lean Leader

The
Lean Leader

A PERSONAL JOURNEY OF TRANSFORMATION

Robert B. Camp

CRC Press
Taylor & Francis Group
Boca Raton   London   New York

CRC Press is an imprint of the
Taylor & Francis Group, an **informa** business

A PRODUCTIVITY PRESS BOOK

CRC Press
Taylor & Francis Group
6000 Broken Sound Parkway NW, Suite 300
Boca Raton, FL 33487-2742

© 2015 by Robert B. Camp
CRC Press is an imprint of Taylor & Francis Group, an Informa business

No claim to original U.S. Government works

Printed on acid-free paper
Version Date: 20141009

International Standard Book Number-13: 978-1-4987-0075-7 (Paperback)

Library of Congress Cataloging-in-Publication Data

Camp, Robert B.
 The lean leader : a personal journey of transformation / Robert B. Camp.
  pages cm
 Includes bibliographical references and index.
 ISBN 978-1-4987-0075-7 (pbk. : alk. paper) 1. Lean manufacturing. 2. Leadership. 3. Organizational change. 4. Industrial management. I. Title.

TS155.C2327 2015
658--dc23   2014039014

Visit the Taylor & Francis Web site at
http://www.taylorandfrancis.com

and the CRC Press Web site at
http://www.crcpress.com

Contents

Preface

Why read this book? Answer: Without Lean Leaders, there can be no Lean.

If you want to be a Lean organization, your leaders must lead using Lean principles.

For almost two-thirds of my career as a Lean practitioner, I could never understand why most transformations failed. As I practiced, read and wrote, the answer was gradually revealed. Lean isn't just about eliminating waste, or saving money, or even focusing on the customer. You can do those things all day long and never be Lean. ***Lean is a way of leading.***

Put another way, until the top of an organization fully embraces Lean, the rest of organization can never be Lean. It doesn't matter how much they spend on consultants; how many Kaizens they conduct; what number of grandmaster, double shot, hold the foam, triple Black Belts they possess. That's all just noise.

Without Lean leaders, there can be no sustainable Lean.

My blog often tackles the nuts and bolts of *leading* a Lean transformation, but my books continue to use the novel format for two reasons. The first is that the characters get to describe concepts in ways that cold definitions just can't. The novel gives the definitions context.

The other reason for using novels is that most of us learn better from stories rather than textbooks. Both can convey the same information, but the novel "puts flesh" on the sterile words as the characters apply them to real conditions.

This book is intended for executives. Not that others can't benefit from it, but the people who really need to *lead* Lean are the folks at the top of organizations.

Please, if you're an executive, don't let the novel format or the short length of this book put you off. It contains powerful insights found in few other places.

The characters of *The Lean Leader: A Personal Journey of Transformation* struggle in their jobs just as you do. They face real crises and what seem to be unreasonable deadlines.

Another thing to take into account: you can't lead a Lean business in the same way as you would a non-Lean one. You'll see what happens as Don, the central character of this story, grapples with that fact.

As the story progresses, you'll watch Don and his subordinates all transform from what is known as *Command & Control* leaders—those who make decisions and bark orders—to a more Socratic method of leading.

As they come to realize that the folks they employ are the real experts of their professions, Don's staff learns to do more asking than telling. They come to realize that their greatest skill needs to be coaching great performance from their people.

They also learn the difference between managing and leading. It becomes apparent that they've spent their careers managing, while calling it leadership.

Slowly, Don and his staff change their behavior to become better leaders. To do that, they begin to shed the decision-making tasks that have cluttered their days. They learn, instead, to delegate those decisions to employees who are closer to the action.

Making decisions closer to the source of the problem causes their organization to be more nimble, to respond to crises and market changes much faster.

Then they *learn to look over the horizon*, to spend more time seeking to understand what the future of their industry and its technology hold, and how they can best position Friedman Electronics, their firm, to be better prepared for that future.

In one of Don's most vulnerable moments, his boss tells him he's not doing his job unless he's building bench. At the time, Don doesn't even know what that means, but as the book concludes, he has begun the mastery of the skill.

All in all, *The Lean Leader* turns out to be someone most of us would enjoy being and, what makes it even better, Lean leaders create organizations that thrive socially, as well as financially.

I feel confident you'll enjoy reading this book.

About the Author

A graduate of the US Military Academy, West Point, I began my career in one of this country's foremost schools of leadership. Yet, like most, I gained my real impression of the skill by watching leaders. I came to realize that leadership is a precious gift that far too many fail to recognize.

In the mid-1980s, as Americans began to watch markets they had created be penetrated, then dominated, by Japan, I started reading the literature trickling out of that country. Each new book pointed to a new technique.

"Statistical process control (SPC) is how they're doing it!" proclaimed some. "It's quality circles," countered others. Still others claimed it was just-in-time (JIT) manufacturing. As it turned out, none was right; all were right.

Over time, as a more complete picture formed, I learned that it was the combination of those tools that gave the Japanese their considerable edge. Still, the picture was not quite complete, and it wasn't until the late 1990s that the world came to realize that the tools alone would never make an organization Lean, because their results couldn't be sustained.

By then, I had made Lean a full-time profession. Over and over I transformed organizations. Some succeeded, but most failed. I tried to make heads or tails of the circumstances that made the difference. The answer, I discovered, was right under my nose all the time. The difference was *leadership.*

I learned that many have been content to manage and call it leadership. Management, however, isn't leadership. In fact, it's a far cry from there. Management faces backward, analyzing yesterday's data and perfecting it today. It's an extremely valuable skill, and much needed, but it's not leadership. Leadership, as I see it, is the ability to look forward, over the horizon and through the haze of battle, to define, then communicate, a new course of action and compel others to follow.

I have had the good fortune to work for some of this country's biggest and most successful organizations. Throughout my career, I've been afforded the privilege of working under great men and women who have taught me both in word and deed. Now a full-time consultant, I was the vice president of operations for a company making specialty apparel for the healthcare industry when I penned this novel.

1

A Retrospective

Jim thumbed unseeingly through the airline magazine. Twenty-five months ago he'd accepted the VP of Operations position and he hadn't been home much since. His wife, Bridget, had overseen the move from their old home into their new one with the efficiency of a NASA mission control manager. She'd also overseen the preparation of their two teens, Joyce and Jim Jr., for the move to their new high school.

Jim, by contrast, had spent most of that time on an airplane visiting the plants now under his control. He marveled again at how he'd gotten this job. He'd actually been a plant manager before this and had been fired by his predecessor. Without his knowing, Jim's subordinates had staged an impressive campaign to win him favor with the board of directors and, shortly after he'd been let go, the board had let the old VP Ops go and rehired Jim into that position.

You might ask why. The answer was that Jim had built a team of subordinates that acted with a near-perfect alignment with one another. They had also improved quality and on-time delivery, while reducing cost to the point that they'd become the best operating plant in the Friedman Electronics Corporation. Of course, that's only part of the reason. The bigger part was that Jim had shown uncommonly good leadership. Because of that, his subordinates had given him their unwavering loyalty. As so often happens with good leaders, Jim had no idea how his employees felt about him until they came to his defense.

Now, Jim always made a point to visit with his old management team whenever he visited the plant. After a rocky start under his successor, the team had resumed their trend of success.

Oblivious to the fact that they'd been successful because of him, Jim continued to feel great pride in watching his old team succeed. He hadn't seen them coalescing around his leadership, but simply banding together to achieve a series of common goals. Although that, too, was true, it had been Jim who had led them to identify the goals in the first place, and Jim who had guided them through the process of achieving them, step by painful step. That was Jim: quiet, humble, introspective, caring.

This flight was taking Jim to visit the worst performing plant in the chain. He was going there to meet Frank, the consultant who had helped him to become the best performing plant manager in the Friedman organization. When he'd been promoted, Jim had rehired Frank to lead the rest of his plants through similar transformations. Some plants had embraced the new style easily; others fought it. Jim had already had to fire several plant managers and he feared this trip would lead to the firing of yet another.

But let's not get ahead of ourselves.

2

Charleston

Don was a Southern gentleman. After a drink or two he might remind you that it was by act of Congress, because he graduated from the US Military Academy in the late 1960s, when graduating cadets were commissioned as both "Officers *and* Gentlemen." But the truth was that Don exhibited—no, maybe "exuded" was the better word—Southern charm.

Don, formally Donald Benjamin Spears the Fifth, could trace his ancestry back to England. Like most Southern aristocrats, the bulk of his family wealth had long since been dissipated and all Don owned were the family plantation home and the 40 acres on which it stood. As an only child, the house and property had come to him through inheritance. His income paid for its upkeep and Don had learned from his parents that, if you cared for something, it lasted much longer than if you didn't, so he took meticulous care of the house and leased the land to pay the taxes. Don's country club membership had come to him the same way as the property. His grandfather had been one of the early investors and not only secured a lifetime membership, but ensured that it would be passed to his son and every heir thereafter. In short, in the town of Charleston, Don was a presence, what many would call a Good Ole Boy, or more derisively, a GOB.

The same was true at Friedman's Charleston plant. Don had risen quickly to a senior management position, largely through patronage, but he'd held his own in each position. When the time came for Don to become plant manager, he had become one of the best in the entire Friedman organization. He'd held that distinction until recently, when, after two years of employing Lean, his peers were starting to eclipse his performance.

Don's management style had come to him in much the same way as his property and memberships: through inheritance. He'd grown up in a household with servants, and the autocratic management style that working with servants required. Don had been led to believe that his authority was the result of good breeding and higher intelligence. As a result, he operated with total self-confidence that his dictates were the best for the organization.

When Jim had assumed the role of VP Ops, he'd visited all the plants in the Friedman chain. He'd been to the Charleston plant during that tour, but at the time, it had been operating above the norm. When he'd arrived this time, he'd asked Don to convene all of his direct reports. After his factory tour, Jim had requested a visit to Don's "Wall."

During his tenure as a plant manager, Jim had been coached to create a single place where the plant's metrics were posted no less frequently than monthly. His team of direct reports had referred to this public spot as "the Wall," because it occupied a wall in the wide central corridor leading to the plant. The title had stuck. Now each plant had one.

As Jim arrived at Don's Wall, he noted immediately that there were only four graphs on the wall and that the graphs only contained the current month's data. There was no board on which the charts were posted. They'd been unceremoniously taped to a wall without so much as a document protector. Jim formed an immediate impression, and it wasn't good.

"Tell me about this," Jim asked one of Don's managers. The man stared at him blankly. Don immediately volunteered that this was "the Wall."

Jim again turned to the manager he'd first addressed and, pointing to the chart labeled "On-Time Delivery," asked the man to explain it. Again, the man stared at him unsure of what to say. Jim turned to a second manager, but before he could ask the same question, Don pushed his way through the knot of managers and began to explain the charts. Jim courteously listened to Don's explanation of the first chart and then held up his hand to impede any further explanations.

Pointing to a second chart, he asked another manager what it meant. The man immediately looked to Don, who started to offer the answer, but Jim again stopped him with a raised hand. "What does this chart mean?" Jim asked again.

"I guess it's talking about quality," the man said.

"Good," Jim replied, "what specifically is it telling us?"

The man stammered, now looking to Don for help. As Don opened his mouth, Jim again raised his hand, never taking his eyes off the manager of whom he'd asked the question. "Go on," he encouraged. "Take a guess."

"Well..." the manager was now visibly perspiring, "... I guess it's saying that we're doing pretty good." There were supportive smiles in the crowd.

"Why do you say that?" asked Jim.

"Well," the manager responded, feeling more confident, "because the point is over 90%."

Jim nodded his head, encouraging the man. Then he turned to the group and said, "There are approximately 87,000 airline flights in the United States every day. If the airline industry operated at 90% defect-free maintenance, how many aircraft could we expect to fall out of the sky each day?" He waited for an answer. No one made eye contact with him.

"OK, I'll give you the answer. We would expect 10% failures, or as many as 870 airplanes to have serious problems. Some of those would fall out of the sky every day. If that were true, how many of you would fly anywhere?"

His question was met with dead silence.

Jim continued, "There are about 20 million surgeries, requiring general anesthesia, performed in this country each year. Anyone here had surgery in the last year? Any of you have a family member who had surgery in the last year?"

Two hands went up.

"If doctors operated at 90% defect free, how many patients would die on the operating table?" Jim waited.

Silence.

"So let's do the math. 20,000,000 times 10% (the number of defective surgeries) would be how much?" A hand was raised near him. "Yes?"

"Two million?" replied the manager.

"Good," Jim complimented him. Now turning back to the two men who had family members with surgeries Jim asked, "If there was a 1 in 10 chance that the physician could mess up your loved one, permanently disfigure them, leave them incapacitated, or even kill them, would either of you let your family member go under the knife for anything but life-or-death reasons?"

Both men wagged their heads "No."

"So," Jim continued, "how is the Charleston plant doing?" He let the question hang a second, then asked, "how many line items do you ship a month?"

There was no answer.

"Who is your production control manager?" he asked.

A man raised his hand.

"How many line items do you ship a month?" Jim asked again.

The man looked confused and shot a glance at Don.

Reasserting control of the situation, Don interjected, "Good conversation, Jim. What say we move it into the conference room where we can all sit down and chat." With that, he moved his team away from the wall and Jim.

"Not necessary," Jim stated. "How about just you and I meet in the conference room?"

"Well," Don began, "if it's just the two of us, let's meet in my office."

Jim could tell Don was trying to move onto more familiar terrain where he had the upper hand. Jim wanted to keep the upcoming conversation on neutral ground. "No, thanks," he replied. "I'd prefer the conference room."

Don, who was used to getting his way, gave the signal to his management team to go back to their jobs, and he led Jim to the conference room. Ever the Southern gentleman, Don immediately asked Jim if he'd like coffee or water. Jim declined, but Don gave an order for coffee to his secretary, who had materialized when they'd entered the room.

Again, Don had demonstrated control over the situation and Jim was tiring of it. Don now chitchatted while they both waited for his secretary to arrive with a freshly brewed cup of coffee. When Don received his coffee and the secretary had departed, he tried to continue his earlier dialogue. Jim stopped him.

"Don, I don't have a lot of time and I need you to listen to what I have to say."

Don stared back with a level gaze. In a jovial tone totally unbefitting the moment he said, "OK, Boss, shoot."

"Don, for better than a decade, you've run this plant and have had great results. I congratulate you. However, change is constant and your progress has not been. You now have colleagues who exceed your performance on every metric and that is concerning me."

"I look at your Wall and know it was hastily assembled, with little plan to continue it after my visit ends. That won't do."

"I look at your management team and realize that not one of them is prepared to replace you if you had a medical emergency. I look at the faces of your management team and every one of them is male. That may have been acceptable in 1980, but this is a new era. The population in Charleston is over 50% female. I also note that you only have one person of color on your staff, despite the fact that demographics indicate that the African American population in Charleston is almost 30% and that there is a growing Latino population."

"Don, you'll note, I'm talking with facts and that's exactly what I expect my plant managers and their staffs to do. I want facts. I want graphs. I want goals set and improvements made. Starting next week, I want you to begin e-mailing me your weekly graphs for this plant's safety, quality, on-time delivery, cost, and sales data. I want to see a plan each week that tells me what problems are inhibiting you from achieving each of these, what the root causes of those problems were ascertained to be, what action you intend to take, who you are holding accountable for the improvement, and what completion deadline you've given them."

Jim noticed that Don wasn't taking notes, nor did he have a pen or paper. "Do you have a photographic memory?" Jim asked.

"No, sir."

"Then go get a pen and paper. I'm dead serious about what I'm asking for and about how I want it. I'll wait."

When Don returned, Jim repeated what he'd told him, and then went on to say, "Don, that's just the start. And, let me be blunt. You're now two years behind your peers. For two years, you haven't taken me seriously and that tells me a great deal about our relationship. It says you don't respect me and that is not a good message for me to get."

"So, I'm putting you on short-term suspension. By that I mean you have 90 days to get with the program. If you do, you keep your job. If you don't.... I'm sure I don't have to tell you what you can expect."

Don's face had grown scarlet, but he kept silent. Jim pressed on.

"Now, Don, I'm going to make every resource in the corporation available to you, but I'm holding you accountable for figuring out what you'll need and for asking for it. Ask me directly. That's how I'll know if you're even making an attempt."

"Do you have any questions of me?" Jim concluded.

Don had kept his gaze on Jim throughout the monologue. He cleared his throat and, after a second said, "No, sir."

"All right, Don. I hear that you're a hell of a guy and it would disappoint me to no end if you didn't rise to this challenge. I believe that you've got the makings of a great leader, but you're a long way from there right now. In fact, I give you less than a 50/50 chance of surviving past a year." He stopped to let his statement take its full effect. "Don, I want you to know that I'm pulling for you and will do everything in my power to help." Jim stopped. "Do you believe me?" he asked.

Don didn't answer for long seconds. Jim waited.

"In truth, sir? No I don't."

"Why's that?" Jim asked.

"Well, sir, I think you've already made your mind up to replace me. I believe you're just waiting for me to fail."

"Don, thank you for the courtesy of an honest answer. I admire that more than you know. If you and I are to make it as a management team, we need to have a relationship of trust built on honesty, so I am glad that you didn't hesitate to give me your honest assessment."

"Let me issue another challenge. Test me. Ask me for the resources you believe you'll need to improve your operation and see if you get them. If you ask and I don't deliver, then you have your answer. If you ask and I do deliver, then I want you to ask for help again and again, until you can accept that I want you to succeed. Deal?"

Don stirred his coffee thoughtfully. "Yes, sir."

"Are you ready to ask for anything today?"

Again Don paused. "No, sir." Then he went on, "Let me give it some thought and get back to you."

"Deal," said Jim.

Jim rose, extended his hand and thanked Don for his hospitality. "You've got my number," he said. "Call me directly." And with that, he left the building and headed back to the airport.

3

At Home

Don's marriage to Huntington Esther Porter had taken place at West Point. Exiting the gothic Cadet Chapel through a gauntlet of crossed sabers, the pair had begun life as a military couple.

Honey, as she was known to her friends, understood her role as an officer's wife. Cotillion life had prepared her for white-gloved teas and receiving lines. When it came to being an officer's wife, Honey represented Don as well as any could have asked.

Honey formed quick friendships with the wives of other officers. When Don's first assignment stationed them abroad, Honey often organized "expeditions," as she liked to call them, into fun spots around their assignments. These expeditions became so popular that the wives of more senior officers often joined them.

The bubbly Honey never failed to treat these women with respect and soon won their abiding loyalty. She didn't pander to them, and was unafraid to be herself.

The young Mrs. Spears' goodwill often found Don being invited to social gatherings with officers two and three pay grades above his own. No dummy himself, Don cultivated these relationships. Having an income from his family trust afforded Don the ability to invite his senior officers on hunting and fishing trips that his peers couldn't afford.

Having grown up accompanying his own father on such excursions, this was a milieu in which Don felt comfortable. He didn't ask for anything from his senior officers. His parents had taught him that, "If you treat people right, and you do your job well, human nature will see to it that opportunities come your way." Don did his job well and sure enough, opportunities had come his way.

Ever the gentleman, Don always expressed gratitude for these opportunities and never behaved as if he felt entitled to them, even though the truth was that he did everything necessary to earn them.

From all outward appearances, Don and Honey were a perfect match, and in many ways they actually were. They'd been friends in high school. Their peers just assumed that they were a "meant to be" couple, and without really intending to, they just fell into an easy dating relationship.

Their marriage wasn't without love, but it was far more sedate than those of others they knew. Theirs was more mutual admiration and affection. Sure there was passion, but the core of their relationship was a shared belief that they would dominate any field in which they chose to participate. In large measure, that became a reality.

4

First Request

Jim had gotten back home after midnight Saturday morning. He was exhausted and crawled into bed beside his sleeping wife. His visit to the Charleston facility had left him frustrated and a bit angry.

His first year, he had followed Jim Collins' advice to concentrate on "first who, then what." Jim Collins had written a highly revered book itemizing the practices that organizations exhibited when transforming themselves from being merely good to being industry-leading. The book's title was true to form: *Good to Great*.[*] It had become like an oft-referenced handbook throughout Friedman Electronics.

Besides, he knew he was being watched and that his honeymoon period would be gone in a flash. He had replaced those plant managers who had not gotten on board. It was easy to pick them out, because their numbers never improved and they refused to adopt the new behaviors required of leaders using Lean.

Those were tense and nerve-racking days, when everyone expected him to turn all the manufacturing plants around in short order, as he had his own plant. The difference had been that he knew who worked for him when he ran his own plant. As corporate VP of Operations, he knew his new subordinates only as former peers. Now he had to get to know them much better.

He'd sent Frank, his now full-time consultant, to visit every plant and to work with plant managers to get them all using some of the tools of Lean. Frank had worked with the plant managers and their staffs. Meanwhile, Jim had met with the plant managers as a group and for one-on-one coaching sessions.

---

[*] Collins, J. 2001. *Good to Great: Why Some Companies Make the Leap ... and Others Don't.* New York: Harper Collins.

A handful of the plant managers adapted quickly. They were soon off and running, needing little assistance. The balance of plant managers fell into two groups. One struggled hard to become what Jim required of them. The others just left or actively resisted. The latter were quickly let go.

As Jim continued to develop the plant managers, Frank then worked on their directors of continuous improvement (CI), a new position that Jim had required each of the plant managers to create.

One by one Frank had conducted Kaizen events in each of the facilities, first demonstrating how to run a Kaizen event, then holding the new CI directors responsible for conducting them with his coaching.

The CI directors received lots of training and practical application in the use of all the Lean tools and philosophies, but that wasn't enough to keep a Lean initiative alive. Breathing life into an initiative would take top-down leadership and that was the phase of his mission that Jim was now undertaking.

As with CI directors, leadership behavior had to be demonstrated at the plant manager and staff level. For that reason, Jim was present at each plant's kickoff Kaizen and set the example that his plant managers should do the same. Jim was back on the last day of each Kaizen event he'd kicked off to hear the final reports of the Kaizen teams, and to encourage them to continue on the path to continuous improvement. It had been a grueling schedule and Jim had gotten the flu several times his first year, but he never let that affect his presence.

What frustrated him about his trip to Charleston was that Don's performance had been good enough from the beginning and so it had never caused Jim to scrutinize it. Now, he felt stupid for having missed the fact that Don had been a highly functioning passive resister.

It wasn't just that he'd missed the fact that Don's metrics, although better than his peers, had never changed. It was more the fact that he'd let two years drift by while Don continued to exhibit the same old bad behaviors to his staff. The way Don had achieved his good numbers was totally dependent on his autocratic style of management. Jim had learned long ago that results attained that way were unsustainable.

Those were his thoughts as he drifted off to sleep that Saturday morning. What worried Jim the most was the fact that he was almost certainly going to have to fire Don. Whether Don made it or not, Jim faced a long rebuilding process in the Charleston plant. The mere thought was draining.

Monday arrived along with a frontal system that dumped buckets of rain. Jim had planned a week in the office to catch up on paperwork. He was well into it when his phone rang around 9 a.m.

"Jim, this is Don Spears. Is this a good time to talk?"

Jim stopped what he was doing and refocused his thoughts. "Yes, Don. How's the weather down there?"

"Muggy," came the reply. "Just the way we like it." There was a trace of humor in Don's voice for a second, but it was quickly replaced with a more serious tone. "If I accept your offer of help, what can I expect?"

"Good start," Jim complimented him. "That depends on what you're going to ask for. Let's review the immediate problems I found with your organization.

- Your Wall, or lack thereof
- Your shallow management bench
- The ethnic and gender composition of your leadership team
- The fact that you've blown me off for two years
- The fact that you employ a dictatorial management style

With which of those do you need help?"

There was no hesitation in Don's response. "All of it, Boss." There was no anger in his words. They sounded humble and contrite, genuinely willing to change.

"OK," said Jim, "let's start by sending Frank to your plant. I can't get him there this week." Jim consulted his calendar. "He's already two and a half hours into a Kaizen event in Topeka. So next week is the best I can do. I'll show up on Monday and the three of us can map out a plan to get you back on track, at least as far as Lean is concerned. Then, you and I can talk about plans for altering your leadership style."

"I want to be clear about something, Don. You are a great manager and I'm sure there is no shortage of firms who'd love to get their hands on you. But I'm not looking for managers. I want leaders and we'll talk about the difference on Monday. So, here's the immediate help I will give you. Next Monday Frank and I will show up and begin the process of getting you back on track. Are you good with that?"

"Yes, sir."

"Don, I'm grateful for the signs of respect, but I don't want you to be subservient. I want you to be as strong and firm as ever, yet with a servant leader's heart. We'll talk more about that when I'm there. For now, I'd enjoy it if you'd call me Jim. Can you do that?"

"Yes, s.... Yes, Jim. I'll do my best."

"Do you need anything between now and Monday?" Jim asked.

"I don't think so."

"OK, remember that you owe me graphs by Friday. If you need help or want feedback, just call."

"See you Monday."

5

Change or Die

Jim's 5 a.m. flight was delayed by almost five hours. He was furious. Instead of being there for the 8 a.m. kickoff, he didn't arrive until almost noon. Before meeting with Don, he went directly to the training room where Frank had assembled the team for lunch.

To his pleasure, Don and his managers were all in the room. "That's a good start," he thought to himself. Before he could signal for him to postpone, Frank asked for everyone's attention and introduced Jim.

"Good afternoon, everyone. I hope you'll keep eating while I talk. You have much to accomplish and your time is short. I just want to make two points."

"First, I'm sure Frank has already told you that Kaizen events are part of the methodology that Friedman is using for continuous improvement. I'll talk more about that in the morning, but for now, let me just thank you for being part of this value stream mapping event."

"The second thing I want to say is that we're all changing. Mr. Spears, me, your management team here in Charleston, the executive team at headquarters. We all are having to change. The reason we have to change is that our markets, our customers, and our competitors are changing. When we are content to stay the same, we are actually falling behind. As the dinosaurs learned the hard way, in life, it's change or die."

"Enjoy your meals. We can talk again later in the week. Meanwhile, what you're doing is making our corporation better and helping to secure your own jobs. Great work!"

There was light applause as people returned to their meals. Jim grabbed a plate of food and seated himself at Don's table. "Hope you don't mind me joining you," he said to those already at the table. "How's the meal?"

Two years ago, when Jim was a plant manager, Frank had started Jim's management team down the path toward Lean by conducting a Hoshin

Kanri to develop top-down metrics. Basically, he led them to assess their mission, ask what commitments it made, determine how they'd measure successful attainment of those commitments, and then appoint someone on Jim's staff to be accountable for the continued measurement and improvement of each.

Here in Charleston, Don's team wasn't ready for that. Not by a long shot. Before he launched full-force into a transformation of the Charleston plant, Jim first had to make sure that Don was going to be on the bus for the whole trip. There was no sense developing Don's management team if Don wasn't even going to be there in 90 days. Again, Jim found himself back to the point of "first who, then what."

So, rather than start the transformation of the Charleston plant with a Hoshin Kanri, Jim and Frank had agreed to begin Lean in a whole different place: on the shop floor. That was not the preferred place to start, but until he knew who the management team was going to be, Jim held off top-down metrics.

In every plant, Frank always started his Lean work by developing a value stream map. That was the macro-level tool that allowed everyone to see the entire process and where their problems were.

So, after Jim, Don, and the other senior managers left, Frank had taken a group of mid-level managers and engineers on a plant walk, what had become known in Lean circles as a *Gemba walk*. Gemba, of course, simply meant the actual place where something was happening. If you had a problem in accounting, you didn't go to engineering, or worse to your computer monitor, to understand it. You went to accounting (Gemba) and observed the problem firsthand.

Likewise, if you had a concern about your manufacturing process, you walked the manufacturing value stream. A *value stream*, Frank had taught Jim, was the aggregate of the processes required to complete one product or product line. Frank explained that when walking a value stream, you always walked from the end of the value stream backward; which, in manufacturing's case, started the team in shipping.

Working their way back through the manufacturing process (value stream), the team recorded critical data about each significant process step. The best book Jim had found on the topic of value stream mapping, as this process was called, was *Learning to See** by Mike Rother and John Shook.

---

* Rother, M. and Shook, J. 2003. *Learning to See: Value-Stream Mapping to Create Value and Eliminate Muda*. Cambridge, MA: The Lean Enterprise Institute.

As they walked, Frank and his team began by constructing a block diagram of the value stream on an 11 × 17 inch piece of paper. Each box in the diagram contained the name of the process step and a symbol comprised of an ellipse in an arc. Frank explained that the symbol was used to indicate that people were needed to perform that step. The number to the right of the symbol indicated the number of people required.

A truncated version of the Charleston plant's value stream looked like this:

Back in the conference room, Frank broke the group into two teams. He had each team write the name of the process steps on sticky notes and tape them to a roll of white butcher paper he had taped to the wall. Because the two teams had already created sticky note flowcharts, Frank now had them use cutouts of symbols he'd printed to complete the value stream map.

5 Days

One cutout was an intermittent arrow. Frank explained that the arrows pointed in the direction of flow and went from each process step to the one on its right. This dashed arrow, Frank told the participants, was the *push* symbol. It indicated that the preceding operation processed parts based on a pre-established schedule and pushed them to the next operation, whether the latter was ready to process them or not.

A *triangle*, he explained, was used to indicate inventory. The symbol always went ahead of the process step where the inventory would be used. The amount of inventory was generally represented by the amount of time it would take to consume the inventory. So, in this case, it would take the process five days to consume the inventory.

Below each process square, he had them tape a data box and record the data they'd collected for the process. Of course, he'd already explained the definition of each abbreviation used in the data box, so by now, they were second nature to the participants.

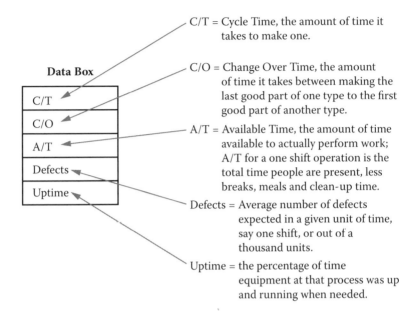

Data Box

C/T
C/O
A/T
Defects
Uptime

C/T = Cycle Time, the amount of time it takes to make one.

C/O = Change Over Time, the amount of time it takes between making the last good part of one type to the first good part of another type.

A/T = Available Time, the amount of time available to actually perform work; A/T for a one shift operation is the total time people are present, less breaks, meals and clean-up time.

Defects = Average number of defects expected in a given unit of time, say one shift, or out of a thousand units.

Uptime = the percentage of time equipment at that process was up and running when needed.

As day 2 drew to an end, there was a heightened sense of excitement in the room. The participants were learning a new skill and they could already see the value of what they'd learned.

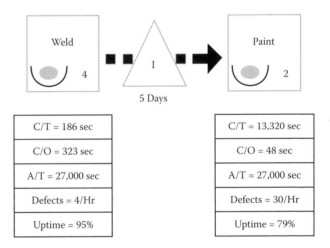

C/T = 186 sec		C/T = 13,320 sec
C/O = 323 sec		C/O = 48 sec
A/T = 27,000 sec		A/T = 27,000 sec
Defects = 4/Hr		Defects = 30/Hr
Uptime = 95%		Uptime = 79%

On day 3, Frank again had the team in the conference room completing the paper doll version of the value stream. The teams completed the data boxes, filling in all the pertinent data gathered earlier.

Then, at the bottom of a value stream map Frank had them draw a new symbol that looked like a square sawtooth. Frank explained that it kept track of value-adding and non-value-adding time of the process. The valley of the sawtooth, Frank went on, was used to record the cycle time (CT) of each process step. That was the time it took an operator to perform the process on one product.

The upper tooth recorded the amount of time the product waited to be worked on. That was the value of the inventory time. At the right side of the sawtooth was a box that had an upper and lower value. The upper value was the total of all *wait*, or *inventory* times. The lower value was the sum of all cycle times. This box worked on the premise that the sum of wait times was the near equivalent of all the time the customer would have to wait for their product. Because *waiting* is a waste, all that time is non-value-adding; hence,

22 Days
11,488 seconds

the sum of all wait time was an approximation of all non-value-adding. The lower value in the box, the sum of all cycle times, was a near approximation of how much time Friedman actually spent working on the customer's job, or, *adding value*.

Dividing the lower value (value-adding) by the upper (non-value-adding) yielded the rough percentage of time that the customer waited during which Friedman was adding value. The higher this value the better, however, in most cases, this value was considerably less than 1%.

22 ~~Days~~ × 24 ~~hours/days~~ × 60 ~~minutes/hour~~ × 60 seconds/~~minute~~ = 1,900,800 seconds

(11,488 seconds/1,900,800 seconds) × 100 = 0.604%

To arrive at that value, Frank had the teams convert the two values into the common denominators (seconds), and divide the sum of cycle time by the sum of lead time. "What this calculation tells us," explained Frank, "is that this value stream only performs value-adding work less than one percent (1%) of the time the customer waits. The rest of the time, the customer is waiting while nothing happens to their order. If you were that customer, Terrance, how would you feel?"

Caught off guard, Terrance responded, "I'd be really pi..." He stopped himself, "really angry," he concluded to the mirth of his classmates. His face was now red as a beet.

Next, Frank had the teams identifying the biggest obstacles at each step in the value stream and the countermeasure that would overcome each. He used the starburst symbol to indicate the countermeasure needed. For instance, it took an average of six hours to change the dies on the big press that punched the motor control cabinet frames out of steel roll stock. Frank pointed out that, because the setup time had been so long, production control had made lot sizes big enough to last a week or more at successive operations. "To get smaller lots, that could be processed faster, the countermeasure was a setup reduction Kaizen event," he told them.

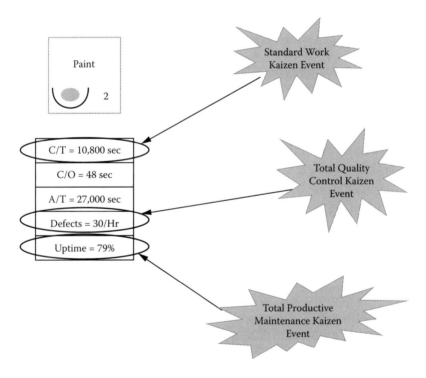

Step by step the team moved through the value stream identifying the problems and their countermeasures. When they had finished, Frank asked, "OK, where should we conduct our first Kaizen event?"

Several suggested the steps with the longest setup time or longest machine downtime. Frank listened patiently, then asked, "OK, if I tackle the setup time of the punch press, way back at the beginning of the value stream, how many more parts can I get out a day?"

That presented the team with a new conundrum. The engineers argued that if you could reduce the setup time by half, you could get twice as many parts through the punch operation. The implication was that you could double the amount of parts out the door.

"Sure," said a manufacturing supervisor, "but they'll just get to the bend process and wait."

Frank let the debate continue for a minute or two before stepping in. Addressing Tom, one of the more vocal engineers, he said, "The value stream map is used to achieve two things: faster throughput and faster cash flow." He asked the engineer to go to the VSM and point to where the company got money. The engineer got up and pointed to the shipping operation.

"Why there?" Frank asked.

"Because that's where we ship product to the customer," came Tom's reply.

"So," Frank challenged him, "does it make any sense to increase capacity in an operation that doesn't result in faster throughput?"

"No, I guess not," said Tom. His voice still sounded unsure.

"OK, then, where is the operation at which improving throughput will actually lead to increased cash flow?" Frank pushed.

Tom studied the VSM then pointed to final assembly. "It looks like it's the restriction closest to the customer," he said.

"I'm not disagreeing," countered Frank, "but why not pack and ship? Isn't it the closest operation to the customer?"

Again Tom consulted the VSM. "It is..." he began hesitantly, "but there is no restriction in those operations." Then the light came on and Tom completed his answer with greater assurance. "From what we can see, they can handle everything coming at them. But, final assembly has a cycle time that's too high."

"How can you tell?" Frank asked.

"Because there is a lot of WIP (work in process) in front of final assembly. Isn't that the first indicator of a problem?" Tom asked.

Frank smiled. "It is, Tom. Again, good analysis."

"So," Frank asked, turning to the whole group, "where will we conduct the first event?"

"Final assembly," came back the chorus of answers.

"But that will barely increase the flow because of all the earlier restrictions in the value stream," one of the engineers argued.

"Right," Frank admitted, "but if I don't open it now, I will have to do it as soon as I open up anything before it. Look," he instructed, "the throughput of final assembly is smaller than the preceding steps. Isn't it logical that as soon as I can get greater volume through the earlier steps, throughput will just be throttled in final assembly? We'll still have the same amount of WIP in the system. It will just be in a different place. And," Frank said, turning back to the engineer, "there will be more labor and materials bound up in the product by that point. Just that much more for which we won't be reimbursed until much later."

There were head bobs as participants started to understand.

"Rule of thumb," Frank instructed. "Always start improvement efforts as close to the customer as possible, then work your way backward through the value stream, opening the throughput pipe as you go. Remember that, especially you engineers. No offense, but most of you were trained to maximize an individual process, but it's the entire value stream that matters. Always ask the question, if I do this, will cash flow any faster? If the answer is no, then this isn't the place to start."

"Now," Frank continued, "every rule of thumb has exceptions. You have to use your head. For instance, if your second operation is throttling your entire operation and you could sell 30% more if it were running in parity with the other processes, you'd tackle it first."

"Same with a supplier whose products have a high defect rate and are causing you a 20% loss of opportunity. You wouldn't wait to work your way back to that. You might not Kaizen it. You might just buy from someone else, but as I said, you've got to use your head."

"New question," Frank stated. "What is the value of this tool?"

"You know what needs to be done," offered one participant.

"And where to do it," suggested another.

"And the order in which to do it," said a third.

Frank then asked, "So, who will get the most from this tool?"

The room was quiet for several seconds. Then someone said, "Well, us."

"OK," responded Frank, "so what will you do with this value stream map?"

Again silence. "We could Kaizen the various operations?" one of the female members queried.

"You don't sound convinced," replied Frank. "Be that as it may, that's a logical impression, which is why I asked in the first place. However, I want to take you through some further logic. Bear with me as I do, please. I promise we'll take a break afterward."

There was a sea of nodding heads, so Frank continued.

"OK, Kendra, you suggest that you and your teammates could conduct Kaizen events, right?"

She nodded sheepishly.

"Do you have authority to shut down a machine that's producing parts?" Frank asked.

Kendra wagged her head and mouthed the word, "No."

"Can you assemble a team of workers to spend five days doing something that doesn't make the plant money right away?"

Again, she shook her head. "No."

"Can you ensure that support organizations such as maintenance and engineering and production control will give you top priority when you need it?"

Embarrassed now, she just stared back at him.

"Kendra, you gave a great and logical answer given what you know. Why else, after all, would we have spent three days working on this tool, right?"

She nodded, a little of her earlier enthusiasm returning.

"Nonetheless, you don't have the level of authority it takes to set a Kaizen event in motion, do you?"

"No, sir."

"Kendra, who does?"

She thought for a second. "Mr. Spears," she replied with new confidence.

Frank pointed at her. "Bingo! Great job, Kendra."

It was always amazing to watch Frank work. He had taken a deflated ego and restored confidence to it in the matter of seconds.

"So, this tool is to be used by Mr. Spears and his staff. They will determine what events to conduct and in what order. They will have to dedicate resources to those events and choose the teams. They'll have to make plans to make enough product ahead of the events to ensure that they don't run the rest of the value stream out of work. We never want to miss our commitments to our customers, right?"

"Right," came the chorus of replies.

Frank stopped. "There's lots to do at this stage of your transformation, but much of it can only begin with your senior leaders."

Pedro raised his hand. "So does that mean that there will come a time when senior leaders won't have to make all the decisions?"

"Very insightful question, Pedro. The answer is not clear-cut, but potentially, yes. I know, that sounded like I was equivocating, and I suppose I was, but here's why. If your leadership team deems it appropriate to establish value stream managers, then those managers could conduct events internal to their own value stream, without having to get direction from Mr. Spears or his staff. Make sense?" Frank asked.

Pedro nodded.

"OK," Frank said, "take your break. Be back in your seats at 10:45 by that clock," he said pointing to the one on the wall behind him. "Great job, everybody."

6

First Who

While Frank instructed the team of value stream mappers, Jim began a series of face-to-face meetings with each of Don's direct reports. He soon found himself depressed by the dearth of management material. Calvin Jones seemed to be the epitome of Don's direct reports. Calvin was Don's operations manager. When asked how he made day-to-day decisions about what products to make, Calvin responded.

"Oh, sir, that's easy. Don tells us every morning what we're to make. He even gives us a list that we're to follow." When Jim asked him what happened when there was a problem and a priority needed to change, Calvin replied, "Oh, that's easy, too. We call Don and he comes right out and tells us what to do. We just do what he says and, sure enough, it all works out just fine."

From all appearances, Don's entire staff looked to him for direction and guidance. None seemed prepared to make—or even be capable of making—independent decisions. Instead, they deferred to Don and talked about his great wisdom and superior intelligence.

It's evident, Jim thought to himself, that Don's idea of a good manager was someone who followed directions well. There's nothing here with which to work, thought Jim. We're going to have to start with a clean slate.

Later, Jim met with Don. "What do you think?" Don asked him. The tone of the question led Jim to realize that Don thought he'd assembled a crack team of subordinates. That made him wonder at the kind of leader Don was. If all he wanted were people who blindly executed instructions, he wasn't going to be of much use to the plant either.

Somehow, and maybe it was Don's military academy training, Jim thought Don had much greater potential than he was exhibiting. Time would tell. Don had fewer than 80 days left to demonstrate what he was made of.

"Don, I'm not one for beating around the bush. There isn't a single member of your staff who could make it as a manager in one of our other organizations."

"Now wait right there," Don responded with indignation. "Those are good people, you're talking about. They've stuck by me through thick and thin."

"Don, we're not talking about quality of character or loyalty to you. We're talking about people capable of independent thinking and decision making. We're talking about people who will be able to build a deep bench and will themselves raise up the next generation of leaders here at Friedman, Charleston."

"But that's what I mean," Don pleaded. "They're good people. They can be trained."

Jim was quiet for a second, then simply stated the obvious. "Don, they might indeed make it in time, but we can't structure an organization around that hope. Your staff's management skills have been allowed to atrophy because you made all the decisions."

"We need to build a new organization. You don't need to lead it. You've got good instincts and have delivered good results, we'll find room for you within the Friedman organization, if you no longer want to preside over the Charleston plant." Jim fell silent. It was Don's turn to speak and Jim was prepared to wait him out. Long seconds passed. Seconds turned into a minute, and still Jim waited.

At last, Don cleared his throat, and in his resonant, patrician voice stated: "My family's lived in Charleston since their land was given to them as a grant by the King of England. Eleven generations of my family are buried in the family plot. I am as much a part of Charleston as it is of me. No, sir, I will not leave. If there's room for me here, I intend to stay."

Jim waited several seconds before responding. He was not surprised by Don's decision. He'd been through this same talk with most of the other plant managers who reported to him and most had made the same decision.

"Don," Jim took a breath and sent off a prayer with it. He knew that what he was about to say needed to be right. "As I told you last time, you have really got to dig in and work hard to stay in your current position. I gave you 90 days 12 days ago and I'm absolutely keeping track of that time."

"If you want to stay in the plant manager position, you're going to have to learn a great deal in a short time; and, you're going to have to exhibit new behaviors and sustain them as long as you're in the position. Am I clear?"

"Yes, sir ... er, Jim."

For some of Don's peers, staying in their jobs had only been a commit-ment to act as if they were reforming while they looked elsewhere. They ultimately left or were terminated. For others, there had been an honest desire to be the leader that was needed, but in the end, some found them-selves unwilling or incapable of the personal changes needed. The remain-der, about 40% of the old plant managers, actually went on to become robust leaders of the new mold. Jim strongly suspected that Don was of the first persuasion. He wanted to be wrong, but strongly doubted he was.

"OK, Don, here's what I want you to do. Next week I want you in our Oakland facility where you'll shadow their plant manager, Jorge Sanchez. You'll tail Jorge and watch how he leads. You'll observe how he leads, which is vastly different from managing, and I want you to pay particular attention to how he deals with his subordinates."

"Each night, I'd like you to prepare a detailed e-mail describing all you've seen. Compare and contrast Jorge's behavior to your own. Finally, I'd like you to tell me what you like about Jorge's style and how it might be adapted to fit your own. Is my expectation clear?"

Don nodded his understanding.

Jim observed Don's demeanor. It was neither dispirited nor that of one giving in to something long enough to find a way out. He liked that. If anything, he read determination in Don's countenance.

Jim continued. "Don, none of the people who currently report to you will be allowed to stay in their current roles. I will make every effort to keep them within the Friedman organization." Don said nothing, but Jim could see that the decision wasn't sitting well with him. "While you're in Oakland, I'm going to begin a search within Friedman for their replace-ments. You'll need to interview each candidate and offer your recommen-dations on which ones I should hire. We still tracking?" Jim asked.

Don hesitated then said, "We are."

"Good."

"On a different note, Don, I wanted to compliment you and your staff for participating in the kickoff of the value stream mapping event. That was good leadership and did not go unnoticed by your employees, or me."

"Thanks, Jim."

"All right, I need to get back to corporate and get this search underway. Any parting words?" Jim asked.

Don thought for a moment before responding. "Jim, you asked for my unfiltered sentiments. That runs against my grain and will take me a little bit to master, but I'd like to say this. I have a suspicion that one of the positions you're going back to search for is mine. If that's the case, then you might as well let me go now." Don stopped and looked Jim square in the eyes. Jim returned his gaze.

"I know the odds are against me," Don continued, "but I intend to emerge triumphant. That's not cocky bravado. I know who I am and I will not sit idly by as I wait to be replaced. I intend to fight with all that's in me, and I intend to win."

"I like that," Jim said with a smile. "I'll hold you to it."

"I hope you will," came Don's reply.

The men shook hands and Jim began his trek back to the airport and to home.

7

Oakland

Don's Sunday afternoon flight had been delayed three times and finally cancelled. He'd had to fly from Charleston to Memphis, Memphis to Houston, and Houston to San Francisco, and then take a taxi to his hotel in Oakland. By the time he got to his room, it was 11 p.m. Pacific time, but 2 a.m. on his body. His meeting with Jorge was at 7 a.m. in the plant lobby.

He laid his clothes out for the following morning. He plugged in his laptop and phone, then quickly checked his e-mail. While he was waiting for his computer to power up, he shot a text message off to his wife, assuring her that he'd gotten in safely and was headed to bed.

He called downstairs and left a 5:30 a.m. wake-up call, changed into pajamas and slipped between the sheets. He was exhausted.

Jorge proved to be an extremely cordial person. He was waiting for Don in the lobby and escorted him to his office.

"How do you like your coffee?" he asked Don.

Jorge's office was small compared to Don's. It lacked the wood paneling and upscale furniture of Don's, too. Looking around the office, Don thought it reminded him more of a kitchen in appearance, with an inexpensive table and six chairs. The walls were adorned with graphs neatly push-pinned to a large cork surface, much akin to a refrigerator covered with kids' art. Although hardly awe-inspiring, Don realized that he felt comfortable in Jorge's office. It felt welcoming.

After he presented Don with his coffee, Jorge said, self-consciously, "Don, I really don't know much about you, but I'd like to learn. In fairness, though, before I pump you for information, let me tell you a little about me."

"I'm a second-generation Mexican American. I'm fully American, and I'm proud of my Mexican heritage. I grew up here in the San Joaquin Valley. My parents first came to this country as migrant workers. Over time, they were naturalized, bought a small plot of land and got work in a local canning factory."

"I have four brothers and sisters, and it was partly because of my entire family working that we were able to buy the house. My parents made every attempt to live off what the kids earned and bank the remainder. We grew up without much of what our peers had, but our parents always made sure we were fed and clothed properly."

"Mom had a penchant for cleanliness. She used to say, 'There is no excuse for being dirty.' She inspected our hands and nails every morning before we left for school. We were never allowed to have our shirts out of our pants, and we always changed out of our school clothes as soon as we got home."

"Homework ..." Jorge said in an almost reverential voice, "... homework had to be finished before we were allowed to do anything; and, Mom had to check it. When my folks learned that my older brother had been given assignments that he hadn't turned in, he was put back in the fields for an entire quarter. The truancy laws weren't as strict then and they were serious about us having a better life."

"I attended school here in California: grammar, high school, and college. I went to work for Friedman as a young engineer and have never left. I have a wife and two children," he turned a photo on his desk around for Don to see, "the rest has been luck, being at the right place at the right time."

Don realized it was his turn. He cleared his throat and began. "I've lived in Charleston all my life. Really, my family's lived there for generations. I have a wife and two children, as well. I went through grammar and high school in Charleston, but went to New York for college."

"Where did you go?" Jorge asked.

"West Point," Don replied.

"Very impressive," stated Jorge.

"I got a degree in engineering, too," Don chuckled. "Funny about that. Friedman seems to hire a lot of us. And," he finished, "I picked up an MBA along the way to my current position."

"Any hobbies?" asked Jorge.

"I enjoy hunting, fishing, and golf," Don replied. "How about you?"

"My kids don't leave much time for hobbies, I'm afraid. Between sports events, music, and Scouts, I'm lucky to be able to read a book." He delivered the last statement with a broad grin.

They finished their coffee over light pleasantries and then Jorge looked at his watch. "I've got my daily staff meeting in ten," he advised. "Let me show you where the restroom is and where you'll be able to set up shop; then we'll reconvene in my office at 7:45."

That night, before Don wrote his e-mail to Jim, he reflected over the day. There were no new problems that he hadn't seen before, but the way they got solved was much different than what Don was accustomed to.

Jim,

In following your instructions to compare and contrast Jorge's behavior to my own, I'm writing my e-mail tonight in a Jorge–Don comparison. Here are my thoughts after my first day.

- Jorge met me in the lobby. By contrast, I have visitors shown in to my office.
- Jorge took time to get to know me. By contrast, I move right to the business at hand.
- Jorge's office was plain—I mentally likened it to a kitchen—and seemed designed to make those visiting at ease. By contrast, my office is large and intimidating with a goal of achieving just that feeling.
- In meetings, Jorge spent most of his time listening. He let his subordinates run the meetings and offered input infrequently. He honestly seemed to be listening to what they were saying. You and I both know that I prefer to run my own meetings and only want direct answers to my direct questions.
- Jorge invited me to his home for supper. I was amazed at how humbly he and his family lived, but they were proud of what they had and shared it freely.
- Their kids joined us for the meal, eaten at a real kitchen table, and much of the conversation revolved around what the kids had done or were doing. Kids led much of the discussion and no one had a cell phone out. If they were texting their friends, it was under the table. Given the amount of attention Jorge and his wife lavished on their kids, I seriously doubt there was any interest in texting.
- No one left the table until the conversation was finished and the table cleared. The kids had to load the dishwasher before they were allowed to engage in any other activity, and no one left the house after supper except Jorge, when he drove me back to the hotel.
- Although their family unit was always the launching point, they made an effort to draw me in to every conversation and wanted to know what my kids were doing.
- If I, on the other hand, were going to feed a visitor, I'd meet them at the club, where my wife would join us after drinks. Kids would eat with a sitter, or with friends. After supper we'd all depart in separate

cars. Even though I had a rental, Jorge insisted that he drive me to his house and take me back to my hotel after the meal.

I apologize, but I'm functioning on very little sleep, so I will end this here. You're right. I'm seeing a different side of life. I'm not sure how those different behaviors affect results, but at least I'm aware of the difference between our two styles.

More tomorrow.

Don

Don watched a few minutes of TV to unwind, then brushed his teeth and went to bed. When he awakened, there was an e-mail from Jim.

Don,

Glad you're getting to know Jorge. He's quite personable and among my best PMs. Don't let his humble origins fool you. Jorge has a double Masters from CalPoly and Stanford. The first in mechanical engineering and the other is an executive MBA.

Jorge's wife, Maria Castanza, who has her PhD in industrial psychology from UC Berkley, is a VP of HR for a pharmaceutical company in Oakland.

Their son, Al (short for Alfonse), is an Eagle Scout, has lettered in basketball all four years in high school and was inducted to the Oakland All-City basketball team three of those years. Their daughter, Marie (short for Maria Castanza), is a three-time invitee to the California All-State Band, as a flutist. Jorge will proudly tell you that she's already been contacted by the Juilliard School of Music.

Though they live humbly, it's not because they have to. You already have a sense of what Jorge makes and his wife actually earns more than him.

My point is that, although Jorge behaves humbly, he has every reason to swagger. His understated behavior is part of the reason his team loves him and would break down the gates of hell to rescue him. See if that knowledge changes your experience today.

I look forward to this evening's note.

Jim

Don had to admit that he was taken aback. Had Jim not given him those insights, he'd never have known and would have figured Jorge for some lucky savant, or the nephew of a corporate officer. He suspected that he might need to rethink what he thought he knew.

The day that followed was a blur. Once again, Jorge met him at the door, but today took him on a factory tour. When they walked through the doors separating the factory from the offices, Don was caught off guard for the second time that day. The factory was clean; not just clean, spotless. Tools, work in process, even waste were organized and neat.

There were colored outlines on the floor for everything. The different colors seemed to correspond to what was inside them. Of course, hazards were marked with yellow and off limits areas were marked with red, but Don had never seen this level of organization before. "How many janitors do you have on staff?" he asked Jorge.

"Only one," came the reply.

"Then, how...?" He didn't complete the sentence. He didn't have to.

Jorge grinned at him and said, "5S is everybody's business."

"5S?" Don asked. "Is that some code?"

Jorge laughed an easy laugh. "You really *are* new to Lean," he said. "Fear not, a week here and you'll be an old hand." He ended the statement with a warm smile and a reassuring hand on Don's shoulder.

There were machines punching and stamping and bending and welding and all the other things that happened in Don's plant. There were people assembling and painting and wiring, just as in Don's plant, but things here just seemed to flow. He thought about that last statement. They really did appear to flow, as easily as water found its way downstream. "What does that really mean?" he asked himself. "Well, the first thing is that they follow a logical progression. Products seemed to move through the production process in the same way they would on a flowchart. They didn't bounce all around the factory from one operation to the next. Every station actually passed the product to the next: hand-to-hand. That's pretty amazing," he concluded.

"Jorge," he asked out loud, "how long have you been at this?" Jorge gave another easy laugh. Don was to find that Jorge responded to most things this way. It was as if he was saying, "It's not all that hard if you have the right attitude."

"We began about five years ago," he stated. "We did it all the hard way, long before Jim became VP of Operations. Now, with Jim's overt support, things are so much easier than when we started."

"Can you tell me what got you started?" Don asked innocently.

Jorge laughed and launched into a well-rehearsed elevator speech. "We started because corporate felt demand was about to increase on this coast, and we were out of space. We were making plans to build a new building when our architect recommended that we use a consultant who could help us design the plant around the way we intended to use it."

"The consultant gave us a brief explanation of Lean and then, for the next three days, he had us talk through how we intended to use the space. He had us create block diagrams of our product's build sequence, then build paper models of where everything would be located and how much room it would need. Ultimately, we built 3D models of the plant. By the time we finished, we knew the exact footprint of not only the building, but the location of every aisle, every piece of equipment, and every tote bin in the facility. It was an amazing process."

"When corporate later killed the new facility, I sent one of my engineers off to get training in something called value stream mapping. Vanessa came back, trained several other engineers, and then briefed me and my staff. My staff and I quickly discovered that we had a tiger by the tail and that the only way to avoid being eaten was to ride it. Here we are," he ended the speech by sweeping his hand across the plant before him.

"So, what's value mapping?" Don asked.

"Value stream mapping?" Jorge asked.

"I guess. Whatever you said a few minutes ago."

"That will take the ability to draw. Let's discuss it when we get back in my office. For now, would you mind if we continued our tour?"

"No. That will be fine, but I really do want to learn. It sounds important if it changed the way you do business."

"Oh, it *is* important, and it *did* change the way we conduct business. You have no idea. I'll promise you this. We'll start our discussion about Lean with value stream mapping; VSM for short, but you'll discover, like I did, that Lean is so much more."

The tour had been eye-opening. It wasn't just that things moved efficiently; it was that no one special seemed to be in charge. Folks running machines were as apt to tell Jorge how they were doing as those in supervisory roles. They all knew their goal for the day as well as for the hour, and could tell you how they were performing against them.

There were display boards throughout the plant. Some showed goals and progress by the hour and others by day, week, or month. Although all were current, it seemed the bigger the board was, the longer the gap between postings. Jorge promised to explain all that later.

Another thing that impressed Don was that Jorge didn't tell people things. He asked questions. In Don's factory, people knew their jobs and stuck to their knitting. Here, people not only knew their job, but the overall plan and how what they did contributed to it. Everywhere, people, as much as materials, seemed to flow in logical sequence from one activity to the next. Don was befuddled. "How could they do that?" he wondered.

At one point, he and Jorge came upon a group in a heated discussion. Jorge didn't jump in. He stayed back and just listened. The group seemed to resolve their differences and the knot started to break up.

"Jimmy," Jorge called out. One of the men turned and came back to address him. "Jimmy, I'd like you to meet Don Spears. He's the plant manager of our Charleston, South Carolina plant."

"Good to meet you, Mr. Spears," Jimmy said, shaking Don's hand. He turned his attention back to Jorge.

"We've been running a trial on the new wire for the TIG welders. Phillipe wanted feedback before he ordered more."

"And?" Jorge prompted.

"The welders don't like it. It puddles too quickly and they've been having problems with blowouts. They want to end the trial early."

"And Phillipe?" Jorge asked.

"There's a lot of money to be saved by switching, so he'd like the trial to run longer, but the men," Jimmy gestured to the welders, "pointed out that they've had to scrap three frames already and rework has cost them an additional 90 minutes between the three of them. They did the math and figure we've already lost whatever savings we might get from the new wire."

"Is it just that it's new material and the welders need to get used to using it?" Jorge asked.

"No way! You know how good these guys are. They could weld diapers on babies. It's just that the wire burns too hot," Jimmy stated.

"Interesting," commented Jorge, furrowing his brow. "And dialing back the amps didn't help?"

"When we did that," replied Jimmy, "the wire stuck to the plate and the wire is so sensitive to amperage that even when you get it dialed in, it still sticks or puddles."

"What's Takt time?" asked Jorge.

"14 minutes and 27 seconds," Jimmy responded. "We've been routinely running closer to 17 minutes since we started using the new wire."

"So, where did it end?" Jorge asked.

"The guys agreed to use it until lunch, but they will change it out before going back to work."

"And you're good with that?" Jorge asked.

"Boss, if we were in a union shop I might think that these guys were intentionally making the wire fail, but I know they're trying their best to make it work. It's just not as good as what we currently buy. I back their decision."

"Good job," Jorge complimented Jimmy.

As Jimmy walked away, Don looked at Jorge. "Was that the supervisor?" he asked.

"Lead welder," replied Jorge.

"You let your welders make decisions like that?" Don asked.

"Who better?" responded Jorge.

"Well, it seems to me that's a management decision, based on the economics of the material. At the bare minimum, engineers should be advising management, based on the data from the trial."

"That's one approach," countered Jorge, "but it takes a lot of time to gather the data, prepare, and make a presentation to management and give them time to make a decision. We've found it faster to train the folks using the material to gather the data and make good decisions."

"I'm a welder by training," Jorge continued. "It's how I earned my way through college, but that's not how I earn my living now. These guys weld 8–10 hours a day. They know if the material is working or not. They can do math. Most of them do trig on a routine basis as part of their job. So why do I need an engineer? And if I involve managers, they're likely to make a decision for all the wrong reasons."

"When buying new equipment or conducting raw material evaluations, the concept we try to work toward is lowest cost of ownership. In a nutshell, that looks not only at the initial cost of the item, but also at any delays that would be created by the new item. It also rolls in the cost of any loss of quality. If the sum of those costs is lower than the current state, then we consider the product favorably. If not, we give it a thumbs down. That decision is simple math. The welders will record their data and meet with the materials group. They'll run the numbers and make the right decision for Friedman."

Don's head was swimming. "Won't managers be angered by hourly workers usurping their authority? I'd think that would ruffle a lot of feathers."

Jorge smiled. He knew exactly what Don was driving at. "We're not perfect," he began, "but we try to get past individual egos and go for team wins. Besides, our managers have a lot on their plates and are only too happy to delegate these kinds of decisions to qualified personnel they've already trained."

"Wait!" Don retorted with an upraised hand, it was a trick he'd learned from Jim. "What could managers have on their plates that could be more important than this?"

"Fair question," came Jorge's response. As usual, it was accompanied by a genial smile. "I've already established that this is a low-level math problem. No rocket science involved, so why tie up a manager doing something this mundane, when they could be working on the next great thing?"

"What do you mean?" asked Don.

"Let's say this would normally be something the engineering and materials managers would decide. What if, instead of working on a low-level problem like this, they were working on a program to identify the materials and processes needed for a new product? Wouldn't that take precedence over deciding what welding wire to use? Whole new product versus shaving a few dollars off welding wire?"

Don knew when he'd been beaten.

"Look," Jorge began, "four or five years ago, you'd have been right. Our managers would have moved heaven and earth to have prevented hourly workers from getting this kind of authority. Back then, it was all about who had the most 'juice.' People worked years to get authority and once they had it, they guarded it jealously. They'd have obstructed Friedman from making a significant gain in the marketplace, if it had meant having to part with their authority."

"Then we began our Lean journey, and people learned that when Friedman won, they won. A Friedman win meant more secure jobs and better personal financial payouts. Holding on to personal power at the expense of the organization made no sense and, little by little, we gave it up. We began working with each other, rather than against each other. We've made a lot of progress, but there's more to go."

"Don, change takes time. People change slowly and if you want them to do things for the right reasons, you've got to repeat the message over and over until they start to believe enough to try. If they try and win, they'll try again. If they try and lose, you've got to keep repeating the message until they try again and again and again, success building on success. Enough philosophy," Jorge said, laughing easily. They continued the tour.

Back in his office, Jorge brewed a pot of coffee while he and Don talked. Mugs of freshly brewed coffee in hand, the two sat at Jorge's table and got to know each other. Don opened up a little about his childhood and how he'd loved to fish as a kid.

"My dad used to have a small boat and we'd go out early Saturday morning for an hour or two. It was the best time of the week. We didn't talk much," Don admitted, "but there was just something about being with him. He always brought a thermos along and we'd drink coffee as we fished. I couldn't have been much more than 10 or 12 when we started. I can't smell coffee and not think of him and fishing. It was a wonderful time."

"Is your dad still alive?" Jorge asked.

"No, he's been gone almost 15 years now. Cancer."

"Mine went the same way," said Jorge. "When you're kids, you can't wait to outgrow your folks and be able to move away, but I can honestly tell you, a week doesn't go by that I don't think about picking up the phone and calling him. If only ..." he concluded wistfully.

When they'd finished their coffee, Jorge went to a whiteboard and began to draw rectangles on it. "Earlier, when we were on the tour, you asked about value stream maps. I'll draw a crude one and then we can discuss its use."

Soon the board was covered with boxes, triangles, arrows, and other symbols. Don wasn't sure what to make of it. Then Jorge started writing in the first rectangles and Don realized that it was a process flow diagram, or flowchart.

"Products flow through the value stream," Jorge began, "from left to right. As they are processed, we capture data on each process step. For example, you can see that the first step, *Punching* takes six seconds a part, that setup takes 45 minutes, that machine *Uptime* is 96%, and that *Available Time* is 27,000 seconds. It's just data, right?"

Don nodded.

"OK," Jorge continued, "if I fill in each of these boxes, I start to get a picture of what's going on in the value stream." Completing the data boxes, Jorge said, "Now, let's do some simple math. What's the longest cycle time?" he asked.

Don scanned the boxes. "Looks like *Bend*," he replied.

"OK," Jorge said, "*Bend*. Let's see. CT = 222 seconds per part. If I divide available time by 222, what do I get?" He slid a pocket calculator across the table to Don.

"121.62," came Don's reply.

"OK, I can make 121.62 parts per shift. By the way, can I ship 62/one hundredths of a part?"

Don snorted at the absurdity of the image.

"I know," Jorge agreed. "Seems like a crazy question to ask, but you'd be amazed how many use that number as their per shift production, when, in fact, they really only make 121 parts per shift. Anyway, let's do the same thing at *Punch*." Don divided available time, 27,000 seconds, by 6 seconds, the cycle time at *Punch*, and said, "4,500."

"So," Jorge concluded, "If I run *Punch* at full throttle, I can make 4,500 parts; then when it gets to *Bend*, what happens?"

Don saw the problem immediately. If Jorge ran *Punch* all out, he'd have punched parts sitting all over the factory. The message was that, until they could increase the throughput of *Bend*, it didn't make sense to punch any more than 122 parts a shift.

"See what I'm driving at?" Jorge asked.

"Sure do," came Don's reply.

"May I ask another question?" Don queried.

"Shoot," said Jorge.

"You used a term, something like *tack time* out on the floor when we were talking to the welder. What is that?"

"Ah," chuckled Jorge. "We can't slip anything past you. The term is Takt time and it represents the speed at which we have to manufacture in order to meet customer demand. You calculate Takt time by dividing available time by the number of pieces you need to make. An example might help."

"Consider a process that has 27,000 seconds of available time and needs to make 500 parts in order to meet customer demand. If we do the math, we divide 27,000 seconds by 500 parts and we get 54 seconds per part. Takt time in this case is 54 seconds."

"Doesn't it matter how long it takes to actually make the part?" asked Don.

"Not to our customer," replied Jorge. "We need to find a way to build them at Takt time in order to meet customer demand. There are a number of techniques to do that, but let's not get off on a tangent. I'm sure Frank will teach you all that."

Returning to the Value Stream Map, Jorge said, "Let me add some more data." He quickly added triangles showing the number of days of work in process (WIP) between each operation. He also showed trucks arriving with raw material at one end, and finished goods departing at the other. "How many products does our customer want?" he asked.

Don looked at the truck on the right, headed to the factory labeled *Customer* and said, "750 per week."

"Can we meet that need?" Jorge asked.

Don seemed puzzled.

"Imagine for a second," Jorge began, "that we only made one product in this value stream. So what is the 'gating' process, the one where I make the fewest items per shift. You identified it earlier as *Bend,* right? Now ask yourself, 'Can we make customer demand of 750 assemblies a week if all I can get through *Bend* is 121 per shift?'"

Don grasped what Jorge was driving at. The factory would have to work a lot of overtime to meet customer demand. Grabbing the calculator, he calculated that they'd have to work an extra nine and a half hours to meet customer demand. "No," came his reply.

"So," Jorge prompted, "if I wanted to eliminate time and a half for the additional parts, what would I have to do at *Bend*?"

"Increase throughput," Don replied without any thought.

"Great!" Jorge shot back, his excitement evident. "See how we use a value stream map? This becomes our road map to identify where we need to expend effort in improving our value stream. My staff and I review ours at least once a month and plan how we'll invest our Lean assets."

"May I interrupt?" Don asked.

"Sure," replied Jorge with a look of concern on his face. 'What's up?"

"What do you mean when you say 'Lean assets?'"

Jorge snorted a laugh. "Yeah," he admitted, "I guess that would be a valuable place to start. Before I do, do you want to take a quick break?"

"You're reading my mind," Don grinned.

"Sure. Come on back when you're ready. We've got almost an hour until lunch. How about Mexican?" he said with a grin.

"Actually," Don said, already at the door, "I kind of had my heart set on a hamburger. Not a chain hamburger. Something with a little more personality."

Jorge laughed. "Got just the place."

When Don returned, Jorge started where they'd ended the previous discussion. "You asked what I meant by 'Lean assets.' To answer that, I need to make sure we're on the same page when we use the word *Lean*. What do you know?"

"Not much," admitted Don. "This has all come on pretty quickly."

"OK, let's start at the beginning."

"Lean is the term we Americans use to define the way Toyota does business. Without getting into a long dissertation on the history of Lean, let's just say that Lean is a business practice that continually looks to remove waste from your product and processes. Waste, as Toyota defines it, is anything that the customer isn't willing to pay for. They refer to such wasteful practices as non-value-adding. So, Lean sets out to eliminate anything that doesn't add value."

"It probably bears noting that Lean does not eliminate people. Just the opposite. Lean views people as the most flexible elements in the business process and, once hired, Toyota works very hard to use them most effectively. I say that, because too many in this country have equated Lean with layoffs. That's because managers in this country have corrupted the message to their own liking." Jorge ended the statement with a smile, but Don could tell this one was different, a sad smile. It was as if Jorge was personally pained by the way others had misused their power.

"Out of one side of their mouth they proclaim that 'our employees are our most important asset,' while out of the other, they are saying 'but we no longer need you.' I could spend the rest of your time here talking about how shortsighted that behavior is, but there's much we need to discuss."

"Before I move on, though, there's a very important point that I need to make. In his book, *The Toyota Way,*[*] Jeffrey Liker documents 14 principles by which Toyota executives live. The first of those principles is: 'Base your management decisions on a long-term philosophy, even at the expense of short-term financial goals.' If you think it through, laying people off is almost always a short-term financial goal. Lean would say, 'If you're losing money, eliminate waste and save.' In a single Kaizen event, you can save hundreds of thousands of dollars, and even more with a Kaikaku event. Just some food for thought. OK, let's get back to Lean assets."

"Any organization trying to transform itself to become Lean, needs to invest effort in achieving that goal. I love Lean and know a lot about it, but, like welding, the transference of Lean knowledge is a full-time job. Although I have a role in that transference, I think I contribute more value by being plant manager. So, the best way to ensure that my organization gets expert Lean advice, is to have someone on my staff take Lean on as a full-time job."

"And not just on my staff. Pretty much anyone who reports to Jim has someone on their staff called the director of continuous improvement, or CI director for short. He or she is the plant manager's Lean expert and the person we hold accountable for developing our value stream map, and for recommending where we will conduct our events."

---

* Liker, J.K. 2004. *The Toyota Way: 14 Management Principles from the World's Greatest Manufacturer.* New York: McGraw-Hill.

"There is a rule of thumb that for every 150 employees, you should have one person in your CI department. That means if you have 450 employees, you should have three: the CI director and two others. The other CI specialists help prepare for and conduct Kaizen events. There's a lot of effort involved and it would work the CI director to death if he or she were the only one doing all of it."

"With one exception, the people in the CI office comprise your Lean assets. The exception is that my staff and I, alternately known as the Lean council, guide the overall direction of our Lean transformation. We assess the VSM and the CI director's recommendations for where to conduct events. Then, once a month, we establish the list of Lean events, as well as their sequence."

"Part of our role as the Lean council is to establish the goals for every Lean event. We never send a Kaizen team in willy-nilly. Each team has a specific list of goals they are to achieve. At the end of each event, they will advise us of their performance against those goals. We set the bar high, but are often amazed that they outperform our expectations."

"Although the CI office prepares for these events, the Lean council communicates the list of those events to the entire organization. We then build ahead in the areas where we'll be shutting down for the week-long event. The last thing we want to do is to disrupt the flow of product to our customers. The CI director takes our plan and executes it, pulling together the teams, assigning the CI specialist, and ensuring that the event we've agreed on comes off without a hitch."

"It's not uncommon for my Lean council to identify who we want as the team leader of the upcoming events. By handpicking the leaders, we ensure that critical people in our organization are deeply exposed to Lean methodologies and philosophies. Typically, the team leader is the person who oversees the process being Kaizened. The Lean council also kicks off every event and, at the end, we return to be briefed by the Kaizen team. This is the briefing at which the team tells us how they performed against the goals we set for them."

"After the event, we hold the supervisor of the area accountable for maintaining and improving on the new performance standards. That process is a great tool to ratchet up performance continually. Forget about staying the same. If we don't continually improve, we fall behind our competitors. You can look around the shop and see evidence of our progress, but we look around and see new opportunities to improve and we go after as many as time and resources allow."

"After each event, the CI director updates the VSM with the new data, then briefs my staff and me. The Lean council reassesses our Lean strategy, revising the sequence of future events as necessary. OK, one last, then we'll go get that hamburger."

"Earlier, when showing you how a VSM worked, I used a situation in which we had a process that needed to improve its throughput. When I did, I kept it in isolation. I didn't tell you what was going on downstream."

"You'll recall that we need to guard against improving the throughput of one area only to have that product sit somewhere else downstream. To avoid that, we generally begin at the end of the process and work our way backward, ensuring that any improvements accelerate the entire value stream, without getting hung up somewhere else. If that's not the case, then we fix the downstream logjam before working on one upstream."

Jorge fell silent, letting Don drink in what he'd just said. It was a surprisingly long silence during which Don re-examined the VSM Jorge had drawn.

"So," Don broke the silence, "if getting more out of *Bend* will cause product to build up at *Weld* or *Paint,* you improve them first, before improving *Bend.* Is that right?"

"Precisely!" Jorge shot back enthusiastically.

"OK," replied Don, "I'm ready for that hamburger."

8

The Spears Estate

When Don left West Point, he had a five-year obligation to the Army. He fulfilled that and, as a courtesy to his commanding officer, agreed to serve a few months extra, while his replacement was located, trained, and posted.

Don had enjoyed his time in the service, and each of his postings, but both he and Honey longed to be back in Charleston. They had gone back once a year, but visits weren't the same. Local friends got married, had children, bought homes, took jobs, and moved. They felt totally out of the loop.

When Don's dad had died, the house and farm passed to Don, his mom having predeceased his dad by a year. By then, the Spears estate was considerably smaller than when his dad had inherited it. Don's dad had made a hobby of farming, mostly using sharecroppers, but he'd never really made any money. In a good year, he was lucky to break even. In bad years, he lost a considerable amount. Like his ancestors, Don's dad had sold off tracts of land so as to pay for ever-escalating living expenses. As he did, the suburbs moved steadily closer. Don's first vow to himself was that the decline of the family holdings ended with him.

Before leaving the Army, Don began looking for a job in Charleston. He landed a position as a manufacturing engineer for Friedman. Over the years, he'd advanced. Keeping his vow, he and Honey lived off his income exclusively. While they did, they reinvested his annuity and the estate began slowly to grow again.

For the first three years, Don rented out all but a few acres of his land to local farmers. He used the small reserve for a vegetable and herb garden. Following the practices of farmers he respected, he was careful about whom he let farm his land, and how they farmed it. For instance, in those first few years, he only allowed alfalfa to be grown on his property. The farmers sold the hay to local dairy and horse farms. The crop fixed nitrogen in the soil. The farmers got cash for their crop; Don got rent for his land; the crops gradually improved the soil. Everyone won.

While others were farming his land, Don spent his time studying. He discovered that some of the best cash crops were annual flower and nursery plants destined for local homeowners. He found a young graduate from a nearby Ag school and hired him to start planting nursery stock.

Don was frugal. Nursery farming was dependent on water, which was no longer cheap. The cost of municipal water led Don to dig wells on his property and to install drip irrigation systems. Understanding that thin-gauge plastic ground cover was as expensive to install as the more expensive woven plastic tarps, he chose the latter. The tarps lasted for years, whereas thin plastic had to be replaced annually. What's more, ground-cover conserved water.

All those investments were expensive, so Don had his nurseryman grow annuals to offset their cost. The first year, Don also installed a greenhouse to get a jump on the following year's growing season. That left the farm income slightly negative. True to his word, Don made up the difference from his salary. The following year, the farm's income went positive and stayed there. The greenhouse allowed him to have spring flowers in grocery and hardware stores in time for the homeowner rush. That earned him top dollar for his products.

The nursery work was intense, and the young nurseryman could only develop five acres a year. That was just fine with Don. Each year he installed another greenhouse and cultivated another five acres of nursery stock, until the whole property was built out. By that time, the farm was running well in the black, despite the fact that he'd increased the salary of the young nurseryman each year.

While Don had occupied himself with work and farm, Honey had seen to the methodical restoration of the old mansion. Don had convinced Honey that, before any interior work should begin, they needed to update the plumbing and electrical systems. At first, Honey had bridled, knowing that the expense would set her behind in decorating for a year or more, but Don's logic prevailed.

"Look," he'd explain, "we know those services are old and will need to be replaced at some point. If we decorate now, and then need to repair or replace those services, it will double the cost." Sometimes she hated his logic, but she understood and agreed. She hired a well-respected architect. He examined the existing systems and set about designing the smartest way to upgrade them.

His designs called for the running of new systems that were redundant to the old. Once the new electrical, water, and sewage systems were run, they'd make final connections, then remove the old systems. Rather than snake pipes and wires through existing walls, the architect designed a new bathroom tower at the rear of the house. That added a laundry room and two bathrooms; one up, one down, something that was much needed in a house built decades before indoor plumbing.

With the design complete, Honey interviewed more than a dozen electrical and plumbing contractors before selecting the ones she wanted. She made it clear that she wanted as little disruption to the house as possible. "We intend to live here throughout the updates," she said. "You find a way to work together and to have things ready to use when you leave at night." Honey Spears was as good a project manager as her husband, and she kept the contractors to their word. After they left each night, she followed up on their work and, if they'd fallen behind, got them to explain how they intended to catch their work back up. Word got out within the contracting community that Honey Spears was no one with whom to trifle.

By the time the kids had come, Honey and Don had fallen into a routine. Honey occupied her days and weekends with friends, a small interior decorating service she'd created, and philanthropic activities. Don, of course, had his job and the farm. They took an annual vacation, but, other than that, had developed largely separate lives.

When Don got home at night, they talked about as much of their day as they believed the other wanted to hear. Supper was usually ready when Don got home, so the two shared a cocktail and news of their days. After supper, they headed off in their separate directions again. There was no question that they were devoted to one another, but details of the other's interior life were slim. They'd long ago established their own domains and didn't need the other to participate in, or provide funds for, their personal activities.

Don's salary paid for the household expenses. His income from the farm was plowed back into the farm. His free time was shared between the farm and his hobbies: golf, fishing, and hunting. Of course, he always accompanied Honey to her philanthropic events.

Honey's income from her interior decorating service had led her to open a small shop in town. The shop sold a mixture of antiques and reproduction accents for homes. It grew quite popular and she spent most of her time and income keeping pace with materials and services. Over time, Honey's business went into the black and actually paid her an income.

The result was that neither Don nor Honey looked to the other for either direction or income.

Of course, Sunday mornings had always been spent at church and Bible study, but each went their separate ways immediately afterwards. If either felt a sense of missing something in their relationship, it wasn't evident.

9

Hoshin and KPI

Back in Oakland, Don was spending day 3 of five with his fellow plant manager, Jorge Sanchez. Over lunch the two discussed children, wives, sports, and pretty much anything unrelated to work. The conversation was convivial and Don found that he and Jorge had much in common. The hamburger turned out to be thick, juicy, and came with all the trimmings. Don left sated.

Seated in Jorge's office, the topic changed back to work.

Jorge began. "Are you familiar with Hoshin planning?" he asked. When Don said "No," Jorge continued. "Jim and Frank are going to take you through this process, but I thought you might enjoy hearing the value of Hoshin planning from a peer."

"Hoshin planning?" Don responded. "What's that all about?"

"It's a great tool and of immense benefit to you," came Jorge's response. "They'll walk you through the mechanics of it; what I want to talk about is why it's so valuable. Frank will take you and your staff through a three or four-day exercise. He'll have you examine your mission and value statements, then he'll ask you what commitments you make in them. Once you've identified those commitments, he'll get you to come up with ways to measure your success in achieving those commitments. Then, he'll have you reduce your performance to graphs and require you to post them publicly."

"Graphs!" Don responded. "Jim's already got me doing some of that. What a royal pain!"

"You'd think so," replied Jorge, "but graphs are really powerful management tools. You'll find that both Jim and Frank will have you use run charts for your metrics. That's because run charts show how you're trending over time. Knowing that, you can tell visually whether the discipline being measured is getting better, staying the same or regressing. You'll appreciate all that more once you've been through the Hoshin. It will be hard work and you'll swear that you'll never use those damned charts, but in short order you'll find how valuable they become in steering your organization's success."

"The first big value is that you and your immediate subordinates will all have a common set of measures of success. When any of you go out to your wall and look at your performance against quality, you'll know immediately how Charleston is doing against that metric."

"The next thing Frank will have you do is determine who among your subordinates will be held accountable for performance against each of those metrics. As an example, you may assign quality to your director of operations. He or she will have to gather the data, update the graph, and account to you for any anomalies in performance. That will drive new behaviors."

"Once you've established the metrics for the Charleston plant, you'll do what Frank calls *cascading*. That's when you assign some component of those same metrics to each member of your staff. That way, you'll have a measure of how they have performed within their discipline."

"You'll be asked to have each staff member own a piece of each of your higher level metrics. So, although the director of operations may be accountable for overall quality, the director of engineering will be accountable for the quality of that department. You may choose that measure to be the number of revisions made to drawings in the previous week, or the number of drawing-related defects found in your product, or both, but somehow, you'll tie the quality of what engineering does back to your overall quality. It's a lot easier than you think, and your staff will definitely have recommendations on how their peers should be measured. You and your staff will want to invest a lot of thought in assigning these intradisciplinary metrics. You'll want to set them so that, when they are achieved, overall success is achieved."

"You said something about alignment," Don interjected.

"Great timing," Jorge complimented him. "Here is one of the huge benefits of top-down metrics. Instead of each of your subordinates measuring her success based on her parochial opinion of how they should be measured; or, measuring it against some arbitrary standard you set for them, they'll all have to agree on what the measures of that discipline's success will be, and that measure will flow directly from the mission! Then they'll have to plot it on a graph and account for any anomalies. That gets everybody on the same page, and everyone's focus becomes centered on mission success, not the success of their own department. Toyota calls that *alignment.* Your entire leadership team is all aligned to achieve the same thing: overall mission success."

"Actually," Jorge continued, "I like learning the meaning of the Japanese words. After all, in Lean, we use Japanese words so commonly, we've anglicized them without ever fully knowing their origin. For example: *Hoshin* literally means "shiny metal," as in the needle of a compass, but is understood to mean direction or even direction-setting policy: strategy. *Kanri* literally means "management." So, putting the two together, we get management direction-setting. So, although we refer to the process as just Hoshin or even policy deployment, what we're talking about is so much deeper. When you conduct a Hoshin and get everybody pulling in the same direction, infighting will dwindle to a faint murmur and your staff will realize how important working together really is to the success of your business. Like I said, Hoshin is a powerful tool!"

"So is that it?" asked Don. "It's just a way to get everyone on my staff to work together?"

"Oh, Don, it's so much more. I've only outlined the first portion. Once your staff is aligned on your mission and measuring their performance against shared, mission-centric goals, you'll have them roll those same metrics down through their entire organization. As you already know, Jim requires his plant managers to post their metrics on a prominent wall in their building, somewhere every employee has to pass. Your entire organization will be able to tell how the company is doing."

"At first most employees won't seem to care, but when you roll the metrics down to their level, they'll start to take notice of your Wall and see how their success or failure affects yours. Every organization will have its own board, and every worker within that organization will need to explain what the charts on the board mean. Soon, your entire organization will be working in harmony. There will be fewer disputes and more synchronization. That," said Jorge, "is huge."

"While you're talking about boards," began Don, "I noticed lots in your factory. They all looked pretty much the same, except for size. What's that all about?"

"You're pretty observant," said Jorge, with a laugh. "Fair question. OK, you have a Wall, so each of your directors has a board. We call them key performance indicator boards, or KPI boards for short."

"Remember we rolled all those top-level metrics down throughout the organization? Well, this is where your subordinates post their organization's performance. You post yours monthly. Your directors post theirs weekly, managers and supervisors daily, and lead people hourly, although their boards will be called production control boards. The boards give you an early indication when things are going sideways. You don't wait until the end of the day, or week, or month to know how everyone is performing."

"Want to take another walk?" Jorge asked.

"Absolutely," Don fired back.

On their way to the factory, Jorge guided Don past his own Wall. On it were a number of graphs, but he singled out the one titled "On-Time Delivery." "If you look at that chart," he began, "you can see that we've been doing pretty well. We've been over 98% on time for months, but in the last two weeks, we've dropped to 96%. Let's see if we can find out why."

They walked to one of the large KPI boards. "This is Eduardo's board," Jorge began. "He's my director of manufacturing. You'd expect his board to have problems with on-time delivery if I'm having them and, sure enough, there it is," Jorge pointed to the drop in manufacturing performance. "So," he continued, "we know the problem seems to be coming from manufacturing, right? Let's go look."

Jorge led them to one, then another, then a third, and finally a fourth board. On the assembly manager's board they could see the drop. "Here's the culprit again," said Jorge. "We're on the trail. Let's keep looking."

They wound through the assembly area to look at the two supervisors' boards. Electrical assembly seemed to be doing fine, but mechanical had a deep dip. "The smoking gun," Jorge pronounced. "Tell you what," he encouraged, "ask someone who looks like a mechanical assembler to tell you about this chart."

Don pointed to someone and Jorge motioned her over to the board. "Consuela," he asked, "can you tell Mr. Spears why your on-time delivery has dropped the last couple of weeks?"

"Bad rivets," she announced. "We've had the manufacturer's technical rep and their salesman living here since the end of last week. For some reason, we got a bad batch and they have been trying to get us good ones. They hand-carried a good batch in yesterday and, so far, knock on wood, they've been working correctly."

"What did you do until you got a solution?" asked Jorge.

"Well," she began, "for the first day we tried drilling them out and replacing them, but they were all bad, so we were drilling and riveting three and four times to get a good product out. Maxine, she's our supervisor," she said for Don's edification, "met with Phillipe and we stopped making any products that used that rivet."

"Show us your quality chart," Jorge asked. Consuela pointed to the chart so labeled on the KPI board. The chart indicated that products had remained an astonishing 99.98% defect free. "We're still only in the five Sigma range, but we'll get there," she said with a confident smile.

"If you'll forgive me, sir," she said addressing Don this time, "we really need to make up for lost ground, so I need to get back to work."

Don nodded and Consuela hurried back to her station.

"Five Sigma?" Don asked.

"Way too early to discuss," Jorge replied. "For the next couple of years, you and your staff need to focus on getting a good appreciation of what Lean can do for you. When you're firing on all eight cylinders, then you can look into other stuff. Just so you know, the actual discipline is referred to as Six Sigma." Switching subjects, Jorge asked, "So, what did you learn?"

"Well, your employees can clearly explain the charts. They can tie the data on the chart to real-life events. They understand when things are within their control and when they're not. They know when to call for help and how not to let the desire to improve one metric ruin another."

"To that last point," interjected Jorge, "these workers know that quality charts are more than ink on paper. They know that they are a measure of Friedman in the eyes of our customers. Oh, and one other," Jorge continued, "we notified the customer as soon as we realized the significance of the problem. We've had conference call updates ever since. We know we're off schedule. They know we're off schedule. There is no shame in explaining why and what we're doing to resolve the problem. We owe them at least that."

"As you indicated, our employees know that getting inferior product out the door to make their OT delivery charts look good is not an option. They didn't need somebody from management to make that decision for them. This is *their* company. This is where they earn their income and they know that if we lose a customer over bad quality, it will have some kind of a consequence on their own job. I suppose it sounds like I'm lecturing," he grinned. "What you're sensing is a passion for what I do and the people for whom I do it; people, including my employees, not just Jim and corporate."

"Let me ask you," Don interrupted. "Did you miss any ship dates?"

"To answer that accurately, we have to look at the Wall, but I can tell you the general answer is 'Yes.'"

"So," Don continued his questioning, "you're prepared to lose a customer over late delivery, but not over poor quality?"

Jorge knew Don was baiting him, but answered anyway. "Absolutely, Don. What good would it have done that customer to get defective product?" Jorge continued. "If you'll humor me, let me ask you some related questions. What would have happened when the defective product got to the customers?"

"It would probably have been caught at their incoming inspection," Don replied.

"Might some have gotten through?" asked Jorge.

"Possibly," Don replied. "But you'd have made your timely delivery, you'd have triggered a payment, and you'd have bought time to find the solution."

"All fair points," Jorge responded. "Next question: what would have happened to the ones caught at incoming?"

"They'd have rejected them and probably sent them back," Don replied.

"OK," began Jorge, "who would have paid the freight on the returns?"

"You would have, I guess," came Don's reply.

"Again, good answer, let me tell you some other costs. The receiving personnel at our customer's would have had to accept a partial order and reject the rest. That would have caused a trackable event in their procurement system and our customer would have had to throw labor at that: phone calls, e-mails, letters, and so on, additional costs to our customer, making it more expensive to deal with Friedman."

"At our end, we'd have to respond to those calls, e-mails and letters. That would have added costs on our ledger and created missed opportunities, because we would have had people responding to a negative situation, rather than pursue a positive one. When those products arrived back here, our quality team would have to write them up, put them in a disputed area, conduct a 100% inspection, write up their findings, create and file paperwork regarding the return, file a written cause and corrective action response with the customer, making a recommendation on how to handle those products; most likely, rework them, and explain what we'd do differently in the future. That would involve additional time of our quality team, manufacturing rework time, re-inspection, and we still haven't gotten to accounting where this problem would create a credit reversal and loads of paperwork on both sides. All those costs could have been avoided," Jorge went on, "by just holding the defective product in the first place."

"Those are all called costs of poor quality on our side, but on the customer's side, they become costs of doing business with Friedman. Those costs get added to our product cost to become part of the overall cost of ownership for Friedman products. If that cost becomes greater than our competition's overall cost of ownership, we will lose that customer; so, getting defective products to the customer on time, was a pyrrhic victory."

"And that's not all," Jorge plowed ahead. "We haven't even dealt with the circumstance of our defective product actually reaching our customer's product. If that happens, we had better hope that they reject it at their final inspection, because if they don't, and the product gets to their customer, all those costs we've just talked about get amplified ten to a hundred times."

"I can only tell you this," Jorge concluded, "Phillipe has been instructed to measure our supplier's *stick rate*. That's the portion of time their products make it all the way into our products. With most of our suppliers, if that rate falls below 99%, we find an alternative supplier; or, if they're a sole supplier, we send in a team to help them get their processes under control. We are rapidly arriving at a place in which we will charge our suppliers if they shut our line down over bad product. In automobile companies, that cost can be over a million dollars a minute. How's that as an incentive to do the right thing?"

Don took a long time to digest what he'd just heard. "You guys think so much deeper about all this than we do. I don't know that we've ever analyzed things from that perspective. I need to think on this some more."

"It's a lot to digest," Jorge agreed. "Why don't we call it a day. Tomorrow be here at 6:50 a.m. and you can participate in our 7 o'clock stand-up meeting."

They walked to the front office where they shook hands and Don headed to the parking lot. His head was swimming.

10

Meetings as a Form of Communication

Jim,

I feel like I've been put through a blender.

Jorge just spent six hours teaching me about Hoshin planning and KPI boards. Much of that time was spent in his plant where I got a real education.

If your purpose in sending me here was to show me that I have a long way to grow, your mission has been accomplished. I understand that you already know what I'm about to tell you, but I'm giving you a summation of what I've learned.

Hoshin: Amazing. To think that a management team could work through all those issues together and develop a common set of metrics and goals. I'm definitely excited to get back to participate in that.

KPI Boards: Flowing top-level metrics down to all levels of the organization in such a visual manner now makes a lot of sense. At one point today, Jorge asked one of his hourly workers to explain her supervisor's KPI board. Not only could she do it, she was able to explain the origin of the problem the graph depicted, and the corrective action she and her fellow workers had put in place.

Lowest Cost of Ownership: Jorge started to discuss that with me. What I understood made a lot of sense, but we've never considered such things in Charleston. I need to spend some more time thinking about all of that.

Jorge has invited me to his 7 a.m. "stand-up meeting," whatever that is. Now that I realize how far his thinking is ahead of mine, I am confident I have even more to learn tomorrow.

Thank you for giving me this opportunity.

Don

Don did something he never did: he took a nap. Afterwards he went for a long walk, ate supper and settled into bed early to read a novel. The novel was less to escape than to give his subconscious an opportunity to process what had transpired that day. Despite his nap, he fell asleep early and slept soundly throughout the night.

When he woke he felt a profound respect for both Jorge and Jim. He now understood how easy it would have been for Jim to just let him go. Keeping him was a selfless act for Jim, whose ego must have made firing him seem like the best solution, certainly the most expedient. Even then, Don knew he had to move quickly to adapt to the new world Jim was revealing to him. The alternative was to perish.

Don was at the plant by 6:30. He left his cup of coffee on Jorge's desk and headed to the plant, where he used the extra time to walk around and just watch people without Jorge present. The first thing he realized was that no one behaved any differently. The plant was just as neat. Work flowed just as easily. Workers focused on their work and managers seemed more involved in observing and conversing than directing. "Astonishing!" he thought.

By 6:50 he was in Jorge's office. They walked to a break room that had been transformed into a war room with flip-chart paper taped to the wall every two inches. Each chart seemed to address a different aspect of business. There were charts for sales made, quotes made, shipments made, first pass yield, on-time delivery, and on and on. A steady flow of managers filed into the room and by 6:57 everyone was present.

"OK," Jorge said, bringing the room to a hush. "Sales, lead out."

His director of sales talked to the number of quotes that had been written in the previous 24 hours, the number of sales orders received, as well as their performance against the sales forecast for the month-to-date.

The director of sales was followed by the director of engineering, who spoke to the on-time delivery of new designs against their planned completion. She also discussed the number of drawings that had been corrected in the previous 24 hours. That was a measure of first-pass yield for designs, Don learned.

Engineering was followed by the director of materials, Phillipe, who spoke to his chart for percent on-time of materials delivery and the percent of first-pass yield of incoming materials. He also updated the room on the status of the rivet problem and the steps that the supplier had taken to resolve the problem.

Discipline by discipline, Jorge's staff talked to the various metrics that addressed their health as an organization. The pace was quick. No one took more than five minutes. Most covered their material in two. This was a well-rehearsed routine and, in 30 minutes, everyone was filing back out again.

Don realized what a highly disciplined organization the Oakland plant was. Each person within it knew his role and what the company needed from him. They each shot to exceed those expectations, treating them as minimum requirements. Often, they met those expectations as groups of two or more disciplines. Don couldn't help but be impressed at the willingness of Jorge's staff to work across disciplinary boundaries.

Jorge hung behind after the meeting, asking questions and following up on discussions the meeting had prompted. At the end, he asked Don, "So what did you think?"

"Amazing," was all Don could say.

"In most organizations," Jorge began, "the left hand doesn't know what the right is doing. Marketing doesn't know what engineering has planned, and purchasing doesn't know what accounting is up to. We make it a point to get all the disciplines together every morning to talk about just those things. If we are going to function as a team, we need to communicate as a team. This stand-up meeting gets us all on the same page. It cross-levels information and tasks. When we walk out, we know what all the other team members are doing and how they are performing against goal."

"You didn't see it this morning, but if anyone falls behind, they are required to discuss their recovery plan. With everyone else's performance tied to theirs, folks know that failure to meet their metrics will result in the organization failing as a whole. It happens from time to time, but we don't give in without a fight. And one other thing," he said with his usual grin, "the stand-up meeting didn't always function this well. We had fits for the first couple of months, but as you can see, the effort was well worth it."

Don felt too overwhelmed to ask even a single question. He might as well have been in an operating theater and been expected to ask questions about the surgery he'd just witnessed.

"Come on," Jorge prodded him, "I've got something I want to show you." They left the break room and walked toward engineering. Before they got there, they were met by Latisha. Jorge introduced them, although Don already knew her from the stand-up meeting as the director of engineering. "Latisha is going to show us the rapid prototyping area," Jorge announced.

Latisha led the trio into a section of their machine shop. It had scaled-down versions of most of their pieces of manufacturing equipment. There were also welding and assembly benches, a small paint booth, and a walk-in oven for drying paint.

"Jorge asked me to show you this rapid prototyping cell," Latisha began. "As you can see, they have pretty much everything that manufacturing has, so we don't have to wait to use manufacturing equipment, nor do we hold them up while we do prototyping. It would have been a big outlay to duplicate the manufacturing setup with new equipment, but we've bought most of our equipment used. We can talk about that later, if you'd like."

Latisha pressed on. "What Jorge wanted you to see was the way the process works. As with most new products we get feedback from the field. We have a weekly conference call with our sales staff. They tell us the problems our customers are experiencing. We talk through potential solutions and frequently develop field-expedient fixes on the call, but then we come back here where we duplicate the problem."

"The call is a two-way exchange," she continued. "The folks on this end of the call will discuss designs they're working on to get the opinion of the salesforce. Sales, in turn, will inform us of problems that customers have with our competitors' products and actually give us ideas for countermeasures to be used in our own. We then bring all those ideas to a team that reviews them for applicability and develops changes to our existing products, or develops all new products."

"Those ideas go to a select group of my design engineers who create back-of-envelope designs that are later reviewed by the entire design team. We further augment this group with marketing, sales, manufacturing, and materials representatives. This larger group then examines the design from each of their parochial interests and makes recommendations for improvement."

"After the modifications from this concept review meeting are incorporated in the design, the design is brought here and a prototype made. Because we often start with an existing design and 'one off' it, this phase of the design process starts with a well-defined list of materials and a fairly detailed subassembly design. Remember, we're just testing viability of the product."

Don raised his hand indicating he'd like Latisha to stop.

"Yes, sir?" she asked.

"One off? What does that mean?"

"Sorry," she laughed. "We get so invested in our own jargon that we sometimes forget that our vocabulary isn't universally shared."

"'One off' means that, rather than starting a new design from scratch, we've taken an existing design and just modified it. In that process, we'll start with a design we know to be robust and only make changes at the subassembly level. At that point, the product's envelope may be too large or too small, but again, we're just testing the concept. Good question," she concluded. "Ready for me to continue?"

Don nodded. He was finding this rapid prototyping process fascinating.

"The prototypes, and any performance data, are brought back to the larger group, which now includes trusted suppliers. This group examines the prototype from a variety of vantage points and proposes any design tweaks deemed necessary. This is our suppliers' opportunity to share with us their ideas about part availability and remaining lifecycle of the parts we intend to use; or, to suggest suitable replacements that may be more robust, or less expensive, or better suited to our use. Their inclusion in the design process has saved us hundreds of thousands of dollars on a single project and millions overall. When we leave this larger group, we typically return to prototype with the proposed changes. If the changes are inconsequential, we may just go directly to the next design stage."

Don raised his hand again. "You involve your suppliers in your design process? Don't you worry about them sharing your proprietary information?"

"Not at all," explained Latisha. "The suppliers we include have reached trusted partner status. To hurt us would hurt them as well. As Jim is fond of saying, 'You can't deliver any faster than your slowest supplier; and your quality can't be any better than your worst supplier's quality.' We take that to heart and only team with suppliers who have great quality and delivery statistics."

Jorge took over. "We walk suppliers through several status levels before they get to trusted partner status. At the lowest level, they get the same terms as everybody else, we both maintain all of the accounting and procurement practices, and their product goes through the normal incoming inspection."

"Our relationship with the supplier grows when they deliver stronger results: better delivery, fewer than expected defects. At that point, we give them more favorable terms and a greater percentage of our business. We also reduce the percentage of their products that we inspect at incoming. When their on-time delivery is perfect and they can show us evidence of statistical process control (SPC) being used in their key processes, we ask them to supply their control charts and we allow them to bypass incoming inspection completely. We call this level of our relationship 'Dock to Stock,' because their product bypasses our dock and is delivered directly to our production line. We may even pay them a slightly higher price per part, if they will also stock the parts in our racks, or deliver them in ready-to-use condition on recyclable racks."

"At this point, we dispense with invoices and pay them directly from their bills of lading. We may even let them walk our floor several times a week and determine what's moving and ensure that their products stay between min and max levels of supply. This is the level at which we also assign them trusted partner status."

"Did that answer your question?" Latisha asked. Don nodded.

Latisha continued with her original discussion of rapid prototyping. "When a concept design is in hand, a series of drawings are made, a Bills of Material (BOM) created, parts costed, manufacturing processes tested on real manufacturing lines, and feedback taken. When we have 98% confidence in the data coming from the design, we're ready to move to the manufacturing phase. Finally, drawings are approved and filed, parts are ordered, standard work is developed and the product is launched."

"Let me interject something," Jorge stated. "Around the same time as the prototype is run through manufacturing, something called a 3P is done. 3P is a combination of standard work and product costing. The 3Ps have been interpreted in a variety of ways. We refer to it as preproduction planning."

"As Latisha has said, we will take the 'one off' design and build it. We conduct time studies of all known processes and also of the newly developed ones. We identify problem areas that we will Kaizen and identify *key characteristics*. Key characteristics are points in the process where quality could drift, so target values are determined and incorporated in our specifications. As products go through these processes, the key characteristics will be measured continuously throughout the build process. By the time we go into production, we have a much better idea of our actual costs and where we fit against our market's price point."

"Even with all this preparation, the move from prototype to production is not without its problems, but there is no casting around or wringing of hands. We know what the product will cost, how long it will take to make it, who will do what to produce it, and how a successful first part should look and perform."

Don was floored. This was so beyond what he was used to, it was as foreign to him as discussing the theory of intergalactic space travel, but this wasn't theory. These people were actually doing it. For the first time, Don could see why Jim didn't feel his staff was up to the tasks ahead. He was beginning to believe the same thing.

Jorge read his look. "Don," he began, "what you're seeing takes years to achieve. We're not showing you to gloat or show off. This is our chance to show a colleague what is achievable. You'll go back and start the process. If you would like us to, we'll invite your people here, or we'll send our people there. Later, you'll perfect things that we'll come to you to see. This is a team after all. Our best results come when we are all performing to our strengths, and learning from each other. Right?"

The more he learned, the more respectful Don became, and the more right he realized Jim was. He felt a bit ashamed for his earlier arrogance. He now had no idea what Jim saw in him, or why Jim had chosen to keep him on. He felt very humbled and his e-mail to Jim that night admitted as much.

11

New World Order: First Who

When Don arrived back in the Charleston plant on Monday he was met with a cup of coffee and a line of people clamoring to speak with him.

"Mr. Spears," said Sue Winters, his secretary, "Mr. Spears, they have let most of your staff go. Everyone is scared that they're next."

The coals of transformation that Don had carefully fanned all weekend were instantly quenched. He was on his feet and headed to find Jim. It was one thing to realize his staff was no longer who was needed, and another to let them go while he was out of the plant. His advice hadn't even been sought about who to keep or who to let go. Don felt righteous anger as he went in search of Jim.

Finding Jim was harder than he would have thought, and the longer he had to look, the more agitated he got. When he finally found Jim, it was in the conference room right next to his own office. By then, Don had worked himself into a state. "What the hell?" he demanded of Jim who was sitting in the conference room with a well-dressed, middle-aged, African American woman.

Jim paused only a beat, then said, "Don, this is Mrs. Jacqueline Highsmith. She's interviewing for your materials manager position. I'll be finished in about 15, if you want to talk."

"No, Jim," Don proclaimed, "I want to talk *now*. You had no right to fire my staff and begin interviewing without me. This is my organization and I will damn well be the one who decides who reports to me." Don knew he was out of line, but this was craziness. How dare Jim!

Jim made apologies to Mrs. Highsmith and motioned Don into his adjoining office.

Fully wound up, Don began again, but Jim simply raised his hand to mid-chest, palm facing outward. The signal was clear. It said, "Stop."

The act infuriated Don. He continued his tirade, getting louder and more agitated with each word.

Jim slammed his palm against the wall and startled Don into momentary silence. "Sit down and shut up!" Jim seethed under his breath.

Don sat behind the hand-carved mahogany desk of which he was so proud. He expected Jim to lay into him, but Jim had turned his back and was totally silent.

Don thought of filling the silence, but the fevered state into which he had worked himself had passed, and now he didn't know what to say.

When Jim finally turned he said, "Don, let's get something straight. This isn't *your* organization. This organization is the property of Friedman Electronics. You have been here as a caretaker with the specific role of producing the products that Friedman designs in the quantities it requires and at a cost that is at or below that of your colleague Friedman companies. In that role, you have failed. You have not achieved any of those objectives."

"Moreover, in the course of pursuing your job, you are to live by the Friedman mission and values. That's your role. Let me make one other point. You serve at the pleasure of Friedman. Your title is not something handed from father to son. It is earned, and, I'll say it again, you have *not* earned it. Am I clear?"

Don sat silent, his anger replaced by dread. He felt cold all of a sudden, and his hands were clammy.

"Silence won't work here, Don. I need to know if you understand that:

A. This is not *your* position.
B. That in recent months you have not met the requirements of this position and are currently on probation.
C. That you serve at the pleasure of Friedman Electronics and its representatives.
D. That outbursts like the one you displayed will not be tolerated ever again.
E. That you had surrounded yourself with a team of underqualified personnel who were incapable of performing their job functions, and that each has been given the opportunity to step backward to a position for which they were qualified, or be let go.
F. As you seem incapable of choosing the correct personnel, I will do so for you.

"Now," Jim asked again, "is all that clear?"

Don lowered his head and shook it in the affirmative.

"Not good enough, Don. You had plenty of words a second ago. I need to *hear* your response. Is that clear?" Jim asked again.

"Yes, Jim."

"Don," Jim stated in a lower voice, "I won't tolerate you behaving that way ever again. I've put you on probation rather than firing you. I kept you because I saw promise. Don't make me regret that decision." The last thing Jim needed here in Charleston was a lap dog, but before proceeding, Don needed to understand the rules of the road. Jim made a mental note to sit down with Don later in the day to make both those points clear. "All right," Jim continued, "I had intended to include you in Mrs. Highsmith's interview, but that boat's sailed. Instead, I want you sit in here and write your thoughts about where you'd like to take this plant and where you'd like to start. I'll be back after the interview and we can talk."

"Oh," Jim said turning, "I'd like you to start by writing out the Friedman mission and values statements." With that, Jim re-entered the conference room and made his apologies to Mrs. Highsmith.

Don tried to write the mission from memory and, after several minutes of struggling, realized he hadn't committed it to memory. He fired up his computer and looked it up online.

"Note to self," he thought, "if they are so important, everyone should have the mission and values committed to memory. That," he thought, "is really at the center of what we do, or at least it should be." He grudgingly cast a thought of admiration toward Jim. Clearly, Jim had already gotten it. He lingered on the thought of mission and values a second longer. "Perhaps," he ruminated, "what is needed is a preprinted card with the mission on one side and the values on the other." He scribbled himself a quick note.

Mission and values copied into a Word document, he began listing the places that he knew from heart needed to be Kaizened. "What else?" he wondered. "Oh yeah, we'd need to do a value stream map, too." Before he could write that last thought, Jim entered his office.

Jim sat in one of the two chairs across from Don's desk. "Why don't you join me over here?" Jim asked, motioning to the second seat.

"May I finish my last thought?" Don asked.

Jim nodded.

Finished, Don printed two copies of the document he'd been working on, then moved to the second seat.

"OK," Jim suggested, "walk me through it."

Don began reading the list. Jim stopped him after each entry and asked him to explain his rationale. He offered no feedback.

When Don had finished, Jim asked him to prioritize the list. Don pulled a pen from his pocket and scribbled numbers next to each entry and handed his copy to Jim. Jim scanned the list and handed it back. "In Lean," Jim began, "there is a definite order of progression. The absolute first is senior management's commitment to using Lean to transform their organization. I'm fully committed. Are you?"

Without hesitating, Don assured Jim that he was.

"You know the story about the hen's and the pig's commitment to breakfast?" Jim asked.

"The chicken contributes an egg; the pig loses a leg," responded Don.

"So, you a chicken or a pig?" Jim asked.

"Pig," Don replied, without hesitation.

"OK," Jim continued. "Next step is to ensure that your entire support team—your direct reports—are all either on board, or predisposed to be. Your original staff, although good people, couldn't even spell Lean; moreover, they didn't even fully understand their roles in the old regime. That is why they are no longer in those roles."

"Just so you know, we offered each of them an opportunity to stay on in a lesser capacity. A few realized that they were unqualified to do their jobs anywhere else and chose to stay on. Most, however, decided to take the severance package we offered them. We could have legally put them out on the street, but Friedman doesn't operate that way. We care for our own, and we did so in this case. You'll be paying their severance for several months into the future."

"Back to your staff. For the time being, I've brought in second-tier personnel from other Friedman organizations, all of whom are in line for a promotion. These people are top-notch practitioners of their trade. You'll get a chance to interview them along with others from outside Friedman. You and I will make the final decision. I also brought in the HR director from my old plant. She's been screening resumes, conducting phone interviews, and setting up face-to-face interviews for a week. For the next several days you and I will be co-interviewing your future staff. Are you up for that?"

Don thought for a long while. "I don't think I can. I don't know what you're looking for."

"OK," Jim came back, "that's an honest answer and I'm grateful, but what we're looking for is their knowledge and attitude. To get through

Flo's interview process, they have already demonstrated that they have the skills and knowledge, at least on paper. After the interviews, we'll go back to references and see if their skills are what they say they are."

"However, when I talk about attitude, I'm talking about happiness. Unhappy people think of themselves as victims. Victims are never responsible for things that go wrong. You can't get them to be accountable, and you need people who will own their discipline. They need to be willing to say, 'Yup, that happened on my watch and that makes me responsible.' Worse, unhappy people often pull those around them down into unhappiness. They do that by creating drama, gossiping, and spreading rumors. Unhappy people tear at the fabric of a team and are unwelcome. That's true whether they are new employees or old ones. It's not a business's job to make people happy. It's not a business's job to change unhappy people into happy ones. That's the individual's job."

"It is a business's job to identify unhappy employees, explain that their behavior is unwelcome and challenge them to change. If they don't, there is no longer a place for them on your staff. They need to go. Period. So, one of the things we'll be looking for is attitude. The other thing we're looking for is a strong work ethic. You can't teach that, either. Maybe you could if the person is young and malleable enough, but not at the age of the staff you're looking for. You need people who instinctively know what needs to be done, and who will do what it takes to see that it happens; not because you're standing over them, but because it's their job and they take pride in what they do."

"It's going to take us some time to find the right people. Meanwhile, work has to continue. As I said, to get us through the interview and hiring process, I've brought trained leaders in from various Friedman plants to act as your surrogate staff, until your own staff is formed. So, back on task, you and I are going to be watching and listening for attitude and a strong work ethic. Can you do that?"

"If I understand you correctly, I believe I can," Don replied.

"Good. After each interview, we'll compare notes to see what each of us thought about the candidate. I'm going to ask that, until you get the rhythm of the interview down, you let me conduct the interview. After that, you'll do the interviewing and I'll ask the odd question or two."

"You on board?" Jim asked.

"I am," Don replied.

"That's good, because the next interview is in 10 minutes."

The balance of the morning was spent in interviews. Flo, Jim's HR director from his old plant, had done an excellent job of arranging candidates. All had a can-do attitude and all were qualified to do the job. Jim explained that the rest was chemistry: did the candidate blend easily with the new staff and with Don? "One thing I want to make clear, Don, and this may be a problem for you; we are looking for people who will have your back, but who will also challenge you. We're not talking insubordination. We're looking for people who know their jobs better than you do and who will unabashedly offer you the benefit of their expertise, even when it contradicts what you think. In the end, you'll need to forge a common path forward, but having a counterposing view can be extremely helpful."

"We are absolutely not looking for 'Yes-men and women.' If this organization is going to succeed, it will be because it has exceptional leadership in every role. From this point on, you and your staff will become the people to whom the rest of the organization is going to look for direction. You won't need to personally have all the answers, but you'll have to know where to get them. You'll all need to be willing to listen to input from those in subordinate positions."

"Now," said Jim, taking a deep breath, "let me address this very issue of having staff members who are willing to take a counterposing position with me. I believe that hearing those views makes me stronger as a leader, but there's a way to do it. As I said, this isn't insubordination. It's respectful disagreement backed by fact. Do you understand the difference?" Jim asked.

Don looked down. "I do," he said.

"What was your behavior earlier?" Jim asked.

Don was quiet for a second, then said, "Insubordination."

"Good, we agree," said Jim. "Don, I need to rely on your strong belief in yourself, your plant, and your people. I need to be able to trust that you will offer your passion, without making it accusatory or defiant. I know that's a narrow path, but good leaders need to walk it. You are former military, can you do it?" he asked again.

"Yes, sir."

"OK, let's get back to the task of looking for your staff, shall we?"

Don nodded with a sense of relief. He had dreaded that he was about to be fired, but even though he hadn't, Jim's point had not been lost on him.

"Normally," began Jim, "we'd look inside your organization for replacements, but there is no bench here. We look inside first because insiders are already part of the culture. They already live the mission and vision of the organization, but, like I said, there was no one here prepared to step into the vacancies. Now," Jim looked Don straight in the eyes, "three years from now, that had better not be the case. The folks you are hiring have got to be constantly building their own organization. They'll do that through delegation of responsibility, by holding their people accountable for meeting their goals and consistently coaching. They'll need to do more, but all in good time. What you need to know now is that I will hold you responsible for the strength of your team. If it's strong, it's because of you. If it's weak, it's because of you. I won't accept the latter. We clear?"

Once again, Jim had turned what had started as a very bad interaction into a teachable moment. Don made no mention of what Jim had done, but he was very aware of how close he'd come to losing what he was now realizing was a spectacular boss. He promised himself that, from this point on, he would accord Jim the respect he clearly deserved.

Don and Jim interviewed all week, except for half an hour each morning when Jim held a daily staff meeting. Don was amazed at the respect Jim commanded from this staff and how well they worked together. Jim had explained before the first such meeting that he would conduct them the first week so that Don could get a sense of what was important and the flow of a good meeting.

Don was impressed at how quickly Jim got his finger on the pulse of the organization and their efforts. Meetings never lasted over 30 minutes, but at the end, every discipline had given its update and left with its directions for the day. He was also amazed by how rapidly the meeting moved and how it resulted in clear direction that would be the subject of the follow-up at the next morning's meeting. It was as if the people in the room shared a common sense of direction and purpose. Each knew his or her own job and led with that strength, but they worked in harmony, never against each other.

Later, Don would learn that they'd all been through the Hoshin Kanri process and all had been part of a team that had developed the direction and metrics for their own plants. What they did in these meetings is what they did in similar meetings in their own organizations. It was amazing. There were six people here from five different plants, but they worked as one team and behaved as if they'd always worked together. Don was truly impressed.

After almost two weeks, the interviewing ended and offers were made. One by one, the new staff arrived and began the job of assimilating into the life of the Charleston plant. Jim and Don had agreed to hire three of the Friedman temps, but kept the entire temporary staff in place for a full week after their replacements arrived.

Each new person was paired with his counterpart from the other plants and shadowed him throughout the day. They were asked to observe and ask lots of questions. "This," they were told, "is the general way we'll expect you to lead. You'll have your own style, but the things you'll do will be those you're watching during this transition. Within the next month, you'll visit the plant of your mentor and spend a week with him. Your mentor has been instructed to take your calls and answer your questions. The worst thing you can do will be to squander that opportunity."

Those who were already Friedman personnel were assigned a sponsor from the Charleston plant. These sponsors answered plant and Charleston-specific questions. Before leaving, Flo arranged for the new people to meet as a group with a member of the Chamber of Commerce. They were given briefings on local lore, customs, and housing areas they would find better than others. This greatly shortened the personal turmoil of the new members.

Jim was the last non-Charleston member to leave. On his last day, he asked Don to drive him to the airport and the two talked on the way. In all, Jim had spent six weeks in Don's plant. He had only gone home three of those weekends and was often to be found on the plant floor day and night, taking notes and talking to the employees.

If there was a problem somewhere in the plant, Jim would be there in minutes to learn, firsthand, what it was and how the leaders intended to solve it. Rarely did he intervene, but he asked to be apprised of the results of their actions. Others watched Jim's behavior and gradually began to emulate it.

During the transition of the plant back to Don, Jim would meet Don's staff every morning, before the staff meeting, at the Wall. Here they would review the events of the previous 24 hours. Below each chart there was an action log. This log was used to record what corrective action was planned, who was accountable for seeing it through, and when the action was due to be complete. In the far right-hand column was a place to record notes and comments. As the new staff had soon learned, you did not want to miss a commitment on this action log.

On the way to the airport, Jim asked Don, "So, what have you learned? What is different from your old way?"

Don was silent for several seconds. If Jim found the silence uncomfortable, Don wasn't aware of it. Finally Don began. "Well, the first thing I see is that your style of leadership is far more distributive. I mean, you don't attempt to do everything yourself, or hold yourself up as the expert in every circumstance. That's a big departure for me. I've also learned that you expect people to solve their own problems and to take calculated risks. You don't expect them to be right or to succeed all the time, and even seem to expect failure, but you always require them to be accountable for their actions and to report what they learned."

Don was silent for a few more seconds before he began again. "Staff meetings are used to recount the trends from the previous 24 hours, good and bad, and to quickly formulate new plans. I've noticed that the staff from other Friedman plants all arrived at the staff meeting prepared with solutions, and not just for the retelling of problems. I've noticed that you use a Socratic method of teaching, rarely issuing directives or mandates, but asking that problems be discussed and interpreted, before a new plan is formulated. I've learned that you give people a lot of latitude," he concluded and then fell silent. Don turned to Jim and said, "I know you have given *me* a lot of latitude. I know that, at times, I didn't deserve it and I want you to know I'm grateful."

Jim was silent for what seemed like an eternity, then asked, "Is there anything to be learned from that?"

"I'm sure there is," Don acknowledged, "but I'd be grateful if you would explain."

"Fair enough," Jim answered, but then took a minute or more to gather his thoughts.

"Don, if I were to describe a colt as having 'good bones,' would you know what I meant?"

"That it had the potential to grow into a better than average horse?" Don replied.

"Exactly," said Jim. "Don, you've got good leadership bones. I see in you great potential and don't want to deprive Friedman of such a leader. Let me ask you, though, if you want to convert a colt with potential into a great horse, what do you have to do?"

Don thought for a second. "Train it. Break it of bad habits."

Neither man said any more until they arrived at the airport. Before getting out, Jim turned in his seat to face the driver. "Don, there's a lot of training ahead. Are you up for it?"

"I think so," replied the other man.

"That's honest," said Jim. "OK, I'm going to send Frank back. He's going to spend a week with you and your new staff. In that week, he's going to help you complete a Hoshin Kanri: a strategy deployment session, if you'd prefer. When you finish, you'll have established your own metrics and will have developed a plan to roll them throughout your entire organization. I'll be back again in a week or two and we can talk about the experience. Meanwhile, I'm available by phone if you need help." Jim extended his hand. "Good luck."

He opened his door, grabbed his bags from the backseat and slapped the top of the roof with an open palm. Don would not see Jim again until he returned with Frank.

12

Cascading Metrics and KPIs

Don really liked Frank. He was a straightshooter and very clear about all his expectations. What Don liked most, however, was the result of the Hoshin Kanri. Now, for the first time, he understood the Wall, and why Jorge had been so positive about the Hoshin planning process. He'd also learned that there were two components to a Hoshin process: top-down metrics and a three-to-five–year strategy. This week, they were only going to develop their top-down metrics.

After he and his staff had struggled with the mission statement, agreeing on the explicit and implied commitments it made, they had then established metrics to measure their progress against each commitment. These *metrics,* or key performance indicators (KPIs), became the charts posted on the Wall. But that wasn't all. Each metric had to have a single point of accountability, a single person (or, as Frank called them, "a single belly button") who would own the performance of that metric against an agreed-upon goal.

When they had completed the top-level metrics, they began tailoring each to every discipline in their subordinate organization. They'd started with manufacturing, as they were accustomed to holding manufacturing accountable to metrics. Identifying the KPIs for them had been easy. They were the tried and true: safety, quality, delivery, cost, and now, 5S. What was harder was choosing KPIs for support organizations. Frank gave an example.

"Suppose," he began, "that one of your top-level KPIs was 100% customer satisfaction. How might a KPI like that affect human resources?" Frank went to the board and wrote: "100% customer satisfaction." Next he wrote eight disciplines: manufacturing, materials, engineering, information technology (IT), marketing, finance, accounting, and HR. "OK," he asked, "what do customers want from manufacturing that will lead to the achievement of this metric?"

Cheryl, Don's director of manufacturing, raised her hand. Frank nodded at her. "They want 100% on-time delivery and 100% quality."

"Excellent!" replied Frank. He wrote on the board:

Manufacturing =
- 100% On time
- 100% Defect free (zero defects)

"Note," he said, "those aren't the only metrics for manufacturing, just the ones that will lead to achieving the overall objective of 100% customer satisfaction."

"All right," Frank continued, "what do customers want from materials?" There was silence. This time Francine, Don's materials director, raised her hand. Frank pointed at her. "Francine?"

"Customers don't want anything from us," she said. "Everything they get, they get from manufacturing."

"Really?" Frank asked. "What is Cheryl's relationship to you?" Frank pursued.

Francine thought for a second then said, "She's my peer."

"Admittedly," agreed Frank, "but in terms of the function you perform, what relationship do you and Cheryl have?"

"Oh, I get it," Francine said, slapping her forehead. "You're trying to say that Cheryl's organization is my organization's customer. Is that right?"

"That's exactly right," Frank replied. "So, if Cheryl is your customer (we sometimes call internal customers *small 'c'* customers), what does your customer want from you that will lead to the overall achievement of 100% big 'C' customer satisfaction?"

This time Francine was prepared. "They want on-time delivery and defect-free material from our suppliers."

"Excellent," Frank encouraged her. On the board he wrote:

Materials =
- 100% On-time delivery
- 100% Defect free

"Marketing?" Frank asked, turning to Jennifer, Don's sales and marketing director.

"Oh, boy," she said. "'I've been worried about getting asked. Let me think.'"

"Well," Jennifer began after a second, "I would think they would include some measure of market share, plus..." she paused for a long time, "I don't know," she said, giving Frank an awkward grimace.

Frank went to the board again. This time he wrote:

Sales and Marketing =
- XX% of market share

Turning to Jennifer he stated, "Although it may not be totally under your control, isn't one of the factors related to the sale of your product the fact that it provides the lowest cost of ownership?"

"Yes," she agreed.

"And wouldn't customer returns of unsatisfactory products also be something you'd be concerned about as a company?"

"Wait!" Jennifer stopped him. "I can't control either of those. Why should I be the one to be held responsible for them?"

"Ah," Frank chuckled, "good point. Before I answer you, let me ask you a question. When a customer buys a Friedman product, do they think 'I like Jennifer, so I'll buy her product?'"

"Well, no," she admitted.

"So, when a customer buys a Friedman product who do they think about?"

Jennifer looked stumped.

"Ask yourself, do they think of a single person, or a single department?" Frank prodded.

"No," she agreed.

"So, when they buy a Friedman product, who do they think about?"

"Us, I guess," she stated as if a bit confused.

"This isn't hard," Frank stated. "They think of Friedman. They think of the corporate entity named Friedman Electronics. So, if you intend to sell products, don't you care about cost of ownership and customer returns? Aren't those potential obstacles you'll need to overcome in the market? And who else would care more about that than you?"

"I'm starting to get it," replied Jennifer. "It's not just what our organization controls, but what the critical measures are that govern how well we can do our jobs. Is that it?"

"Precisely."

Frank went back to the board and completed the sales and marketing portion of the metrics:

Sales and Marketing =
- XX% of market share
- Lowest cost of ownership
- Zero customer returns

Before moving on, Frank turned to Jennifer and asked, "Can you now see how these metrics are important to you?"

"I can," she admitted.

"OK, Philip," Frank turned his gaze to Don's human resources director, "what should you be responsible for tracking?"

"Boy, Frank, after that last exchange I'm feeling like I might be responsible for global warming and world hunger."

That got a laugh from his colleagues.

"But seriously, I am confident that retention rate will be one of them. Other than that, I'm not sure."

"Who's responsible for training?" asked Frank.

"Oh, yeah," replied Philip. "I guess that would be me, too."

"So," Frank said writing on the board:

Human Resources =
- XX% Employee retention
- Zero skills gaps

"Here's one that takes a bit of intuiting," Frank admitted: "Annual compensation rate based on the cost of living index (CoLI). For now, you'll just have to trust me. The way in which you administer employee pay is about to change and they will, in part, be pegged to the cost of living index. I say 'in part,' because the other way you will reward employees will be with a bonus program based on this plant attaining clearly measurable achievements. If the plant doesn't hit that, employees get the same as they got last year, adjusted for CoLI."

"I suppose a digression here would make sense. One approach to bonuses is only to give them to those people who did something to exceed expectations, but if you think that through, we would only be giving a few people recognition and our goal is to win as a team. So, we want a reward system that will reward team wins. Make sense?" he concluded.

Philip nodded.

Frank completed the table:

Human Resources =
- XX% Employee retention
- Zero skills gaps
- Annual compensation based on CoLI and bonus

"OK, Philip, what is your current employee retention goal?"

"75%," he stated.

Frank turned to Don. "Is that acceptable?"

Don started to answer, but Jim touched his arm. Don stopped and Jim whispered something in his ear. Don seemed to weigh what he'd been told, then answered Frank's question. "Jim tells me it's not," he said with an affable smile.

"Apparently, the rest of Friedman is running between 90% and 93%. Philip," he asked, turning to his HR director, "what are our chances of achieving that goal?"

"Zero to none, Boss. We already blew way past that. We are in a period of churn, what with the recent changes in your staff. I'm sensing unrest and a lot of people are leaving as a result."

"What's achievable?" Don asked.

"This year? More like 65%–75%."

Frank wondered if Jim would intervene, but the latter seemed content to let the dialogue between Don and Philip continue.

"If we accept that what is past is past," began Don, "and we begin measuring as of next month, what's achievable in the rest of the year?"

"That's not up to me," said Philip. "It's really the other directors who will need to answer that question."

Don held up his hand. "Much as I'd like to do a roll call and determine who will stay and go, this isn't the time or place. I do, however, expect each of you to do just that and report back by this time next week. Meanwhile, Philip, I think your goal should be 88% for the rest of this year. Is that doable?"

Philip thought for a second. "Don," Philip began, "that's quite a stretch, but I'm willing to try."

"Understood. Please begin measuring our performance against that value." Turning to Jim, Don said, "With your permission, Jim, this is all new to us. Let us succeed this year and then we'll get more aggressive. Is that OK?"

"More than OK," Jim replied. "I also want to compliment you and Phil on the game of catchball you played in getting to that value."

"Catchball?" Philip asked.

"It's a technique used to work out an acceptable goal between a leader and his or her subordinate," said Jim. "Remember how it works, because you're all going to have to use it as you work out goals for your subordinates."

Jim nodded at Frank indicating that he should take over again.

"So, our top-level HR metrics are:

Human Resources =
- 88% Employee retention
- Zero skills gaps
- Annual compensation based on CoLI and bonus

"Oh, and one more," stated Frank. "You'll want to have 100% of employee engagement studies completed."

"A question?" asked Philip.

Frank nodded at him to proceed.

"Where did employee engagement studies come from?"

"I can't slip anything by you," responded Frank, grinning.

"Let me answer that one," offered Jim. "One of the things that Friedman Electronics has begun is conducting an annual employee engagement study. That study uses a questionnaire to look at the health of your workforce using several key indices of employee engagement and morale. We believe that if an organization really believes the statement, 'Our employees are our most important asset,' then caring about how employees perceive their job and their relationship with the organization becomes critical. Make sense?" Jim concluded.

There was a murmur of assent around the room. Jim nodded to Frank to continue.

"If those are the measures of HR success at the top of the organization," Frank continued, "what will be the metrics at the next layer down, the manager level?"

There were no responses.

"Phil?" Frank asked.

"Well, I'm guessing that 88% retention rate will be on the list, and I guess there would need to be something about skills gaps and employee engagement studies, but I can't see them being worried about compensation at that level."

"Good beginning," Frank complimented him. "Let me help you." He started a new list.

Manager Level Human Resources Metrics =
- 88% Employee retention
- 100% of employees covered by skills matrices and gaps identified
- 100% of employee engagement studies completed

"Clearly," Frank continued, "the lower in the organization we go, the more granular their metrics will become. Moreover, although not always true, the lower you go, the shorter the interval between measurements. So, if we measure monthly at your level, we might measure weekly at the manager level. Supervisors are always measured no less frequently than weekly and lead persons are measured hourly. Let's go down a level and look at supervisors. At that level, the metrics would look something like this:

Supervisor Level Human Resources Metrics =
- 88% Employee retention
- 100% of employees covered by skills matrices and gaps identified
- 100% of employee engagement studies completed
- Rolling 12-month list of retirements or planned resignations identified (24 months for leadership positions)

"Would we have an HR goal at the lead person level?" he asked.

"Unlikely," said Philip.

"Great. Then we've got HR covered. Now," Frank continued. "I'm sure you noted that lower tier metrics aren't all flow downs from the higher levels. That means you'll have to use your noodles determining lower-level metrics. Always ask yourselves, 'What do we need to measure in order to attain the higher level metric?'

"OK," he concluded, "who's ready for a break?" People started moving immediately. "Twenty minutes back here," Frank called after them.

13

It's About the People

The rest of the day had been spent fleshing out the remainder of the metrics for Don's entire organization. For the most part, Frank let the Lean council, those who were Don's direct reports, develop the metrics on their own. He only intervened if they were stumped or going off track. When the day had ended, everyone was exhausted, especially Frank.

"Same place tomorrow, 8:00 a.m.," he'd told them.

With that, they'd filed out, mentally exhausted, but quietly expectant. They had, after all, developed a clear way of knowing how they were doing, and it was tied to their agreed-upon mission.

They began again the following morning. Once the goals for each top-level KPI had been established, and the member of Don's staff accountable for its performance had been identified, the team agreed that each director had to report on progress no less than monthly. If performance fell below goal, the frequency of reporting could increase all the way to daily.

As they had developed subordinate KPIs, they ensured that every person in a management role was accountable for no less than four metrics. Frank had explained that, by continued and frequent review of their subordinate's charts, managers at every level could keep mission success on track.

Like Don's "Wall," each manager was to have a board on which all of their KPIs were posted. Once again, performance against KPIs would be depicted by a run chart, showing both the goal and performance against it. Below each KPI chart was to be a sheet for action items. Every action item had someone from that organization who was accountable for it, and a date by which it was due.

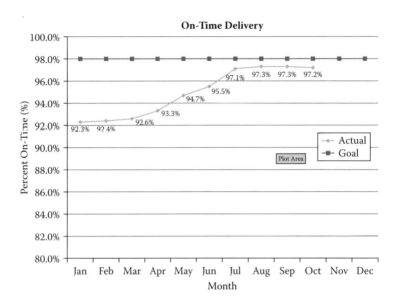

Action Item List

Item#	Date	Problem	Description	Responsible (Name)	Due Date	Status	Comments
1							
2							
3							
4							
5							
6							
7							
8							
9							
10							
11							
12							

"If you show your people you're serious," Frank had told them, "by holding meetings at your KPI board, discussing both good and adverse trends, assigning accountability for corrective actions and then following up—constantly following up—then you'll see performance start to improve."

Frank had worked with Don's continuous improvement (CI) director, Tim, to identify a local supplier that would screen print all the KPI boards. Together with local managers, Tim found prominent locations for each of the boards, identified them on a blueprint of the floor, and then reviewed their proposed locations with Don. Don was pleased with the younger man's work. It was both thorough and timely.

On Jim's next visit, he had asked Don to call his staff together for a two-hour meeting. Everyone was concerned that a VP had come to meet with them. The last time Jim had been here, it had involved the shakeup of the entire Charleston management team. However, they need not have worried.

When he arrived, Jim met with Don and Frank. Jim reviewed his plan for the meeting and asked Don if it was all right for him to run it. Don thanked him for the briefing and for not assuming the right to address his staff. "Not that you wouldn't be within your rights," Don continued, "but you've made me a partner in what's about to happen, and not diminished me to the status of one of my directors by letting me learn at the same time as them. I'm not surprised, Jim, but I'm grateful for being treated with advance knowledge of your plan."

When everyone was seated, Jim thanked everyone for coming and told them that he, Don, and Frank were going to take them through a new exercise intended to improve their leadership skills. With that, he asked, "How are we doing?"

When no one responded, Jim spoke again. "That's a twofold question, isn't it? I'm asking how you and I are doing. I'm also asking how Friedman Electronics' Charleston plant is doing as part of the overall Friedman system. So," Jim pointed to Francine, Don's director of materials, "how are we doing, Francine?"

A brief smile played across Francine's face before she answered. If she was intimidated by Jim she didn't show it. "I think Charleston is doing extremely well," she began. "And, if our return on shareholder equity is any indication, Friedman Electronics is doing well."

"Good," Jim smiled. "Anyone else have a differing opinion?"

No one indicated that he did. "OK," Jim smiled again, "since I have such a timid group, let me ask the next question. How are *WE* doing?" With that, he pointed to himself, then to each person around the table, except Don and Frank.

Don's director of HR raised his hand. "Yes, Philip."

"Well," Philip began confidently, "I'd say we're doing fine."

"What leads you to say that?" asked Jim.

Philip pulled a face, as if he hadn't expected the follow-up question. "Our metrics all seem to be improving. Our Lean council has learned a great deal about Lean and how it affects each of our disciplines. Employees seem happy."

"Funny you should say that," Jim responded. "What makes you think your employees are happy?"

"Uh," Philip had been caught off guard again, "I don't handle as many complaints about managers. I see more smiling faces," he added as an afterthought.

Jim turned his attention to the group. "Let me ask you. How many implemented suggestions have you had in the last month?"

Jim's gaze was met with blank stares. A few looked at each other hoping they had the answer. "Anybody?" Jim asked.

Finally Tim, the CI director, raised his hand. "I had a couple last month," he supplied.

"OK, consider this," said Jim, "and this is really important. We've learned that implemented suggestions are one of the best measures of employee trust. Said a different way, they are a measure of how your employees trust you." He let the statement just sit out there for several seconds. "Now," he asked, "how are we doing?" Again, silence. "Let me make a point, Lean is so much more than KPI boards and metrics and improving. At its core, Lean is about respect for people."

"Think about it. If your suppliers trust you, Francine, they will find ways to partner with you. That means they will work with you to redesign your products making them less expensive and better functioning."

"If your customers trust you, Jennifer," he said, indicating Don's director of sales and marketing, "they will seek your input on development of their next generation of products. And," Jim changed his tone for emphasis, "if your employees trust you, they'll offer their insights on how to improve their processes. What do you think about that?" he asked, turning to Cheryl, Don's director of manufacturing.

"When you put it that way," Cheryl replied, "I guess we're not doing all that well."

"Thank you, Cheryl," Jim said, smiling at her.

Turning to the rest of the directors, he asked, "How many of you have submitted suggestions?"

Two people raised their hands. "How many of those got implemented?" Jim asked.

One of the two raised her hand.

"How's the trust within Don's staff, the Lean council?" Jim asked.

His question was met with silence, but Jim could see his point had registered.

"OK, somebody tell me what point I'm trying to make."

Again Cheryl led off. "You're implying that the level of trust between Don and the Lean council isn't all that high."

"Is that a valid assessment?"

"I don't think so," Cheryl replied, "but if you use implemented suggestions as the measure, then you would certainly be led to that conclusion."

"Look, I know you've had your hands full for the last month," Jim continued, turning to let his glance embrace all the directors. "I'm not finding fault with any of you. You've clearly been doing well. I'm raising the expectations you hold for your people and for yourselves. One of the ways that Toyota went from an obscure, underfunded automobile company to the number one auto company in the world in just 60 years, was their reliance on employee suggestions. Implemented employee suggestions are a critical part of any continuous improvement system."

"Cheryl, not to pick on you, but perhaps you can tell us who are the most experienced welders in your shop."

Without hesitation, Cheryl named three.

"Great," Jim replied. "So, who understands the day-to-day problems with welding better, your welders or an engineer?"

Cheryl sensed a trap and said, "Well, both have their place. I'd go to each of them for different reasons."

Jim smiled. It was obvious that Cheryl was concerned about the message she'd send if she was presumed to be badmouthing engineers in favor of her welders.

"You don't trust me, do you, Cheryl?"

She froze, stunned.

"Let me explain why I said that. You gave me the 'politically correct' answer. You really believe your welders are the people you'd go to, but you're not sure what I'd do if you chose them at the exclusion of the engineers. In short, you don't trust me. You're afraid what I might do to you or someone else, aren't you?"

Again, Jim's question was met by silence. He turned to the rest and said, "If she doesn't trust me, there is no good answer to that question, is there? If she says 'No, Jim, I don't trust you,' I might take retribution on her. If she says 'Yes,' she worries that one of you might betray her real feelings and she'll get in trouble that way. No answer was a good one, so she chose silence."

"OK, we've got a bigger problem here than implemented suggestions, so let's take a sidebar and talk about the issue of trust. Hypothetically, why might Cheryl not trust me?"

There was quiet around the table, then Don said, "Well, she could have heard that you fired my entire staff before she got here and might worry that the wrong answer might get her the same fate."

"Perfect, Don. I was hoping someone would get to that point. Thank you! So, you're all treading on eggshells until I leave, right? That's no good. That makes me being in your plant a cause for fear. That just can't work. We need to be able to trust each other."

"Don," Jim asked, "why was your staff let go?"

Don sat silently for a few seconds, then launched into his answer. Turning to his new staff he said, "I wasn't silent to think of the most 'politically correct' answer. I was collecting my thoughts. Jim's taught me to do that. Here's the answer, as much as it pains me to say it."

"Jim visited this plant, met with me and my staff and rightly concluded that they had all been picked to execute my directions. I did not want pushback or clever answers from my subordinates. I wanted … well, I wanted 'yes people.' When Jim arrived, my numbers were below most of the other Friedman facilities. Jim would have been within his rights to fire me right there, but he didn't. Instead, he sent me to a sister plant for a week, and allowed me to observe the way they operated. One day there, and I saw what Jim saw. I was ashamed of myself."

Don fell silent, then started again. "When I got back, I was told that Jim had let my entire staff go. I flew into a rage and confronted him in front of a woman he'd just started interviewing. Again, Jim would have been within his rights to fire me on the spot, but he did not. He made me do an exercise that taught me how out of sync I was with the rest of Friedman. Then he explained his rationale for letting my people go. None of them were capable of making independent decisions, because: (a) they didn't really know their professions the way you do, and (b) they were chosen to implement my ideas, not have their own."

"I'll also point out that each had been offered the chance to step down into a job they could do, but only a handful actually did. The others took a severance and left. No one was actually fired."

The room fell silent for a second, then Jim picked up the thread.

"Thanks, Don. For the rest of you, there are some important messages here. The first is that, if leaders leave a communication vacuum, the grapevine fills in the blanks. Some things to know about the grapevine:

- The grapevine almost always gets it wrong, and usually does so while putting legitimate leaders in an unfavorable light.
- The grapevine characteristically accentuates the negative.

"If that's true, what's the best thing that leaders can do to cut off grapevine speculation?"

Cheryl raised her hand.

"Trying to get back on my good side?" Jim asked with a smile. It was met by chuckles around the table, including Cheryl's.

"Can't hurt," she said with a smile. This time the crowd laughed.

"I'd say that the best thing leaders can do to prevent gossip is to tell the truth."

"I knew I liked you," Jim said. "Perfect answer. In the world of Lean, we call that *transparency*. That means that we tell the truth and we do it right away, even when the truth is unflattering. So," Jim pushed on, "Do you think you have reason to fear me?"

There were negative headshakes around the table. "OK," Jim concluded, "We've cleared one hurdle, but we're not ready to get back to discussing implemented suggestions. Let me ask you this, how much do you know about each other?" Again, he let silence settle in before continuing.

"Francine, is Cheryl married? Does she have children? What are their names?"

Francine was struck dumb by the intimacy required of the answers. Before she could even attempt to do so, Jim turned to Tim and asked, "Tim, what's the state of health of Francine's parents? Does she have siblings? How are they?"

"Are you getting my point?" Jim asked. "Let me show you what I'm after. If we trust each other, we begin to share our lives—what's really important—with each other. That trust leads to great things in the workplace. Turning to Don he asked. "Don, what's my wife's name?"

"Bridget," Don responded.

"How long have we been married?"

"I think about 18 years," came Don's reply.

"Close, 19."

"Do Bridget and I have any children?"

"Yes, two. Let's see, your eldest is a girl, Joyce, as I recall. And your son is Jim Jr."

"Have we had trouble with either of our children?" Jim pursued.

This time, Don gave Jim a penetrating look before responding. "It's OK," Jim told him.

"Um, Joyce had a problem with drugs," Don said sheepishly.

Jim looked at those around the table. "Frank," he asked, "how bad was Joyce's problem?"

"Serious. When you discovered it, you had her hospitalized and then put her through a rehab program. This May will mark her third year of being clean and sober."

"So let me ask you," Jim addressed the assemblage, "do I trust Don and Frank? And now, by extension, do I trust you? We have got to develop relationships of trust," Jim ran on. "Every one of your employees goes home each night. Is it to an empty apartment? Is it to a pet? Is it to a spouse or a partner? Are they caring for parents, or kids, or grandkids? Until your employees trust you, they'll be reluctant to share their ideas with you. Now, don't get me wrong, simply knowing a little about their personal lives won't make employees trust you, but it often opens the doors if, and this is a huge if, you are sincere."

"Tell me, Kanisha," Jim asked Don's director of engineering, "what do we mean in Friedman's values when we use the word *transparent?*"

Kanisha thought for a moment. "That you can see our motives, that we don't have hidden agendas, that what you see is what you get."

"Nicely put," complimented Jim. "And may I add, that the motives you can see are ones that imply mutual success?"

"Yes, of course," responded Kanisha.

"What's the takeaway, Chris?" Jim asked Don's director of information technology.

"That we should be transparent," suggested Chris.

"And?" asked Jim.

Chris, always taciturn, thought for a second. "That we should get to know our people?" His response was more a question than an answer.

"Half right," Jim replied. "Anyone else want to give it a shot?"

Tim raised his hand. "That we should let our people know who we are?"

"Are you asking me?" Jim responded.

"That we should let our people know who we are," Tim stated more confidently. "That we should be dependable. They should be able to know in advance how we'll respond to most questions, because we'll have made our intentions transparent to them."

"Together with Chris's answer, that's the whole answer. You want to know your people. I mean really get to know them. You should not only know what their skills are, but what their career desires are, as well. It's not bad if they don't have aspirations to be CEO someday," Jim continued. "We need assemblers, and wave solder machine operators, and data entry clerks, and receptionists. If they're comfortable being there, that's not bad, but two things you'll need to know is how to motivate them and what skills they lack to be the best at what they do."

"How do you find out what motivates them?" he asked.

"We talk to them," stated Philip. "We get to know them."

"OK," agreed Jim. "And how will you know what skills they lack?"

Again Philip interjected, "We interview their supervisors."

Jim said, "At the start, that's exactly right. First, supervisors will determine what skills are needed for each position. Next, you'll need to establish a way of objectively determining who can do what, and to what ability. Then you'll create a skills matrix, like we talked about yesterday. The skills matrix will depict what level of proficiency each employee has passed. That matrix will, on one axis, list all the skills required by that supervisor, and on the other axis list all the people under the same supervisor. The boxes formed by the intersection of the lines will then be color coded, using the familiar red, yellow, green color system. Green means the person is fully proficient. Yellow means that they are proficient enough to perform the skill with supervision. Red means they are not proficient and should not perform the skill."

"Now, I began this discussion by qualifying what Philip had said. Supervisors will keep the skills matrix, to begin, but then they will turn them in to human resources who will keep a master. Each month, the supervisor will make provisions to test any employee wishing to test at a higher skill level. If the employee passes, she will advance to the new color and the new sheet will be sent to HR. Any questions?" Jim asked. There were none.

"OK," he began, "I have an assignment. For the next hour, I want this Lean council to develop two things:

1. A list of questions you feel every manager should be able to answer about their direct reports; and,
2. A skills matrix covering those people who report to each of you.

"Are there any questions?"

No one raised his hand and discussion began almost immediately. Jim and Frank stood to leave. Don looked at Jim. "Should I stay?" he asked.

"Up to you. If you think you'd learn something from observing your staff complete their assignment, then by all means, stay. If you're confident that you could do the job yourself, then you're welcome to leave, but I'll be looking for your skills matrix for each of them in an hour," he said with a wink.

"In that case," Don responded, "I'm confident that they will do well on their own. I think I'll go to my office and begin working on my assignment."

With that, the three of them exited the room.

An hour later, Frank and Jim met in Don's office. "Ready?" asked Jim.

"I think so," Don said with a satisfied smile. "Come look."

The three moved into the conference room, where Kanisha was at the board and recording the team's work. She moved with lightning speed as the last of her colleagues completed the names and skills for their skills matrix.

Kanisha gave a quick wave, alerting the rest of the group that Jim, Don, and Frank had entered the room. A few looked at their watches, but they all understood that their time was up.

"So, what have we got?" queried Jim.

People started passing their papers to him. Most had hand-copied what Kanisha had been writing on the board. Manny turned his laptop to show that he'd captured them in Excel. He later confessed to having sent each of his fellow directors copies of their skills matrices through the e-mail system.

Examining each of the matrices, Jim passed them to Don and Frank.

"OK, these look good. What about the things you should know about your employees?"

"Do you mind if I address that?" Philip asked his fellow directors. When no one objected, he pressed on. "We felt we should know the names of significant others in their lives. That included spouses, partners, children, grandchildren, as well as live-in or ailing parents. We also felt we should know what they aspired to accomplish in their careers. You know, were they just here for the paycheck and the medical benefits, or did they aspire to do more with their lives. Beyond their aspirations, we wanted to know why they came to work each day. Were they putting themselves through school, or paying off medical bills, or saving up to adopt, or buy a house, or an engagement ring?"

"How do you plan to do all that?" Jim asked.

There were a few glances around the table before Cheryl spoke. "We felt we should create a 3 x 5 card for each employee, then ask questions and begin to fill them out. When the cards are full, each of us intends to use them as flash cards until we can recite the information for our direct reports from memory."

"Sounds like a good plan," Jim congratulated them. "This won't be done in the open so that it looks like a scavenger hunt, will it?"

"No, sir," Cheryl stated with an impish smile. "We figured you'd put the kibosh to that, so we planned to do it over time and make it appear as if we're doing what we're doing: showing sincere interest in them."

"I should have known better than to worry about that with you guys," Jim said with evident admiration. "OK," he continued, "let's take a break. When you get back, we'll have index cards for each of you and you'll perform the exercise on each other. When you walk out afterward, you'll know about your peers, too. All right, I've got 10:43," Jim said, consulting his watch. "Let's all be back by 11:00."

With that, they all trooped out.

When everyone was back in their seats, Jim handed out 5 x 8 cards. These were printed front and back. As the cards snaked around the table someone exclaimed, "These aren't our questions. They're a lot more detailed."

Someone else asked, "Why didn't you give us these to begin with?"

Jim smiled. "You just learned one of the secrets of deep knowledge. You will have much deeper knowledge of a subject if you struggle to understand it. Another thing that deepens knowledge is what I call *soak time*. That's when you perform a new task, think about it for a while, then prepare to teach it to someone else. You'll discover that you have much deeper insight having given the topic time to soak. So, who wants to start?" Jim asked.

Philip raised his arm at the elbow. "I guess I should," he said. "Will it hurt?" he asked, getting a ripple of nervous laughter.

"One more thing," Jim added, "the person who asks Philip questions will be the last to be asked. Does that offer an incentive to anyone?" he asked with a smile. "Also, the person just asked, will be the next to ask. So, Philip, you will ask the next person and we will go clockwise from the person who asks you. Of course, you'll skip over the one who asks Philip until the end. Clear?" Jim asked, looking around the table. "So, who wants to ask Philip?"

Cheryl raised her hand.

Cheryl began on Question #1. "Where were you born?" she asked.

"Stuttgart, Germany," Philip responded. Turning to the rest of the table he stated, "My dad was career Army. Retired after 32 years." There was clear pride in his statement.

"Question #2. Name your siblings and their relationship to you in birth order."

"Sister, Kathleen, 18 months older; brother, Kevin, two years younger; brother, Bradley, ten years younger. Bradley was a bit of a surprise." That got a laugh.

"Question #3. "Are your parents still alive, and if so, where do they now live?"

"Dad passed in 2002. Mom's alive and lives with Kathleen in Walnut Grove, California. She'll be 85 next year."

"Question #4. Do you have a spouse or a partner? If so, what is his or her name?"

Philip was silent for several beats, then began. "Some of you know, Margie and I got divorced last year." He stopped again. "We'd been married two weeks short of 25 years."

Out of respect, Cheryl waited a few seconds before asking the next question.

"Question #5. Do you have any children, and if so, what are their names and ages?"

Now Philip took long seconds before beginning. When he did, it was with his brimming eyes downcast. "I have two," he said. "Christine, we call her Christy. She'll be 16 in March. Carmen, we call him Chip. He'll be…" Philip's voice trailed off and everyone could tell he was struggling. He started again. "Chip will be 13 in June." His voice cracked on the last statement and Cheryl waited almost a minute while Philip regained his composure. Kanisha, who was seated next to Philip, put her hand on his back and gave it a consoling rub.

Finally, Philip said, "Sorry. I guess that cat's out of the bag. I was given joint custody, but needed to get away from Margie. When I moved here, it meant seeing the kids less frequently. I really miss them." This time, Philip let himself break down and began to sob in their presence. I …," he began, "need to…" He got up and left the room.

Everyone immediately looked to Jim for guidance. He simply held out a raised hand, indicating that they should all wait. There were several minutes of silence before Philip re-entered the room.

After he seated himself, Jim addressed the group. "What you just witnessed is not unusual behavior when you get people to open up about themselves. If they choose to make themselves vulnerable to you, what's the worst thing you can do?"

He waited. Finally, Kanisha raised her hand. "Act like nothing happened?"

Francine was next. "Later share their confidential information."

Tim said, "Remain aloof to their pain."

"All true," Jim stated. "So, if that is how you should not act, how should you act?"

There was silence as those around the table formulated a response. Confident that each had gotten the message, Jim said, "You can continue."

"Question #6. Name your hobbies in order of interest."

"Tennis!" Philip stated with renewed excitement. "Then golf and fishing. Can you tell why I'm in Charleston?" he asked with a grin.

Many around the table chuckled for Philip's benefit, knowing all the same that he was putting on a brave face for their benefit. They played along.

Round the table the questions went. Most gave safe responses, carefully avoiding revealing anything close to what Philip had.

Then it became Cheryl's turn to answer, and Philip's turn to ask the questions.

Question #1. "Where were you born?" he asked.

"In a hospital," she responded, getting an immediate laugh. "OK," she picked up, "in a bed." Again, laughter. Cheryl was used to controlling the crowd. Finally, she admitted, "Rochester, upstate New York."

"Question #2. Name your siblings and their relationship to you in birth order."

"Older brother, William. We call him Will. Younger sister, Hope."

"Question #3. Are your parents still alive, and if so, where do they now live?"

"I never met my real dad, so I can't answer that part. My last stepdad is dead, thank God. There were a couple others, but we never kept track of them. My mom is still alive. She lives with my sister. If you couldn't tell," she played to her audience, "my sister is the *good* one of the family." That, too, got a healthy laugh.

"Question #4. Do you have a spouse or a partner? If so, what is his or her name?"

Cheryl had become flippant. "Well let's see.... Number one, Billy, we lasted about nine months. Number two lasted longer, almost three years. Number three lasted about two years. Not a great track record," she admitted with a smirk. Again, laughter. "No one in the pipeline at the moment. I kind of think I'm not meant to be married."

"Question #5. Do you have any children, and if so, what are their ages?"

"Three," she responded. "Beth, she's 27. Cynthia, we all call her 'Sin,' and the title fits." More laughter. "Let's see, she'd be 23. Then there's Judy. She's my baby and she'll be 22 this May."

"Question #6. Name your hobbies in order of interest."

"Don't really have any," Cheryl stated. "I read when I can and love history, but Judy lives with me, and by the time I get home, she's ready to talk. Not much of a life, I know," she said with what sounded like the first genuine feelings.

That was it. Cheryl had been the last one to answer questions. The process had taken over an hour.

"Let's take a break," Jim said. "Ten, then back in here."

When the council reassembled, Jim asked, "So, what did you learn from the exercise?"

Tim raised his hand. "A lot about what motivates us."

"Explain," said Jim.

"It's hard," Tim admitted, "but the questions we asked seemed to get to the root of who we are and what drives us. I mean," he stammered, "none of the questions got there directly, but when you put them end to end, and fill in the gaps, a picture forms. I don't know," he concluded, "it just seems like I've got a better idea of what gets my colleagues out of bed every morning."

"Anyone want to comment on that?" Jim asked.

"You know," Kanisha began, "you just figure everybody is like you, and then you hear about their lives and you realize that we're all different and that some of us have worked a lot harder to get where we are than others. It gives me a great deal of respect for my colleagues."

"Anyone else?" Jim asked. There were no further comments. "OK," Jim concluded, "do you understand your colleagues a bit better now?" There were nods and murmured assents around the table. "And you think that will help you in working together with them?" More nods and a few yeses. "And you realize that this was all preparatory to getting to know your employees better for the same reason, correct?" More yeses, and more emphatic nods.

"All right," Jim continued, "I don't want to lose sight of the fact that we entered this swamp to drain it, not kill or elude alligators. Our goal had been to discuss the importance of implemented employee suggestions."

"Frank, can you take over?"

"That I can, Jim. Let me assure you, this will take no more than 15 minutes, then, with Don's permission, you'll be free to go."

While Frank began, Jim turned to Don and said, "I hope you know I'm not trying to upstage you by conducting this discussion. It's a topic you and I haven't reviewed, and I had hoped that if I did it in your hearing, we'd kill two birds with one stone. That could give you a several-week jump on the implementation of what we've discussed. You OK with that?"

"Absolutely," responded Don. "As a matter of fact, I am really glad that you conducted it and that I was able just to listen. I learned a lot from them and from you. Thanks."

"OK, before Jim hijacked the program…" Frank paused to let the laughter die down. "Implemented employee suggestions; you know what that means. The question is how do you get your employees to offer them, and how does that topic tie to understanding your employees?"

Jim continued to marvel at how Frank so quickly summarized discussions and segued into the next topic. "Let's start with the last question first," he began. "How does understanding your employees lead to implemented suggestions? The answer is trust. Your people need to trust a couple of things before they'll start volunteering their good ideas. The first thing they'll need to trust is that you will honor their idea by giving it serious consideration; that is, you won't laugh at their idea or ridicule it, or them."

"The second thing they'll need to trust is that you'll do something with their idea and not just let it fall on deaf ears. At bare minimum, you owe it to them to give them feedback. If their idea was accepted, they should be told. If not, they should be told that, too. The feedback you give should include any pointers on what they could do to improve this idea so as to make it acceptable or pointers on how to get their next idea accepted."

"OK, I want to give you an example that I give to every Lean council. I call it 'the probability of getting to yes' example. We'll discuss the moral after I relay the story."

"Picture someone in Cheryl's organization, down working on a machine, and he comes up with a brilliant idea. This idea, if implemented, will save $200,000 a year and only costs $20,000 to implement. That's $180,000 net savings the first year and $200k every year thereafter. Sound like a good idea?"

Frank looked around the room, then turned to Don. "So, Don, what do you think?"

If Don sensed a trap, he didn't let on. "It seems so," he replied.

Frank asked him, "How much authority does the person who made the suggestion have to make the change?" There were knowing smiles shared around the table before Don answered.

"Well, in our organization zero. They'd have to go to their boss with the idea and then take it up the chain."

Still addressing Don, Frank said, "OK, you took probabilities back in high school, let's test your recall. We've already agreed that this individual has no 'yes' authority. That is, he is denied the ability to spend $20,000 and implement the idea himself. That leaves him with two possibilities: 'No' authority—that's a double entendre, by the way—but what I mean is that if he exercises his 'No' authority, he chooses to say 'No' to promoting his idea. The alternative is what I call 'up arrow' authority: the ability and willingness to promote his idea to his boss. You agree?"

"Sounds about right," Don replied.

"OK, we've been left with two events: 'up arrow' and 'no.' Knowing nothing else, what's the probability of each?"

"50%," came Don's response.

"Good," Frank affirmed. "Now, which of the events could lead to the implementation of the idea?"

"The up arrow."

"Right," Don agreed, "and we've already discussed that it only has a 50:50 likelihood of even being discussed with the next-in-line supervisor."

"Now, what authority does the next-in-line supervisor have? Can they authorize the expenditure of $20,000?"

"Not in this plant. In fact, I don't know of any Friedman plant where they could."

"You're not unusual," Frank ran on. "So, that leaves the next-in-line supervisor with only two options, right? And would you agree that each of the options has 50% probability?"

"Yes."

"And what is the probability of getting the up arrow from the next-in-line supervisor?"

"Same. 50%," said Don.

"Very good. Now, remember that to get the first up arrow was a 50:50 probability, but to get the second one, you first had to get the first; so, the probability of the second is the probability of the first, 50% (.50), times the probability of the second 50% (.50). Point five times point five equals .25. In essence, you reduce your chances of getting the $200,000 savings by 50% every time you have to go up a level to get the authority to implement the idea."

"How many levels in your facility?" Frank asked.

"Four."

"Do the math, what's the probability of getting the $200,000 savings here?"

"Zero," Don responded. "I don't have that authority."

"How many more levels would you need to go through before someone did?" Frank asked.

"At least two," Don replied.

"Great! So let's do the math. Every level is 50% or .5 and there are six levels." Frank went to the board and wrote:

$$.5 \times .5 \times .5 \times .5 \times .5 \times .5 = .015625 \times 100 = 1.56\%$$

Frank turned to the rest of the Lean council. "Now, I was kind. I gave every level a 50% chance. What would happen if anyone in that chain had been belittled or berated for promoting an earlier idea? Would the probability of getting approval go up or down?"

"Down," remarked Kanisha, as if a light were coming on for her.

"Imagine," Frank continued, "if the person with the money-saving idea never left the starting gate, if they'd been unsuccessful in a previous attempt and just kept their idea to themselves; exercised their "No" authority. What is the probability of getting the idea implemented?"

"Zero," chimed in Tim.

"Exactly," said Frank. "Now, do you understand why you might not be getting any good ideas? Remember what Don looked for in subordinates in the past? By now, your employees believe their ideas won't go anywhere anyway, so why even try."

There were knowing looks around the table.

Frank was on a roll. "One of the ways to improve your odds is to teach your people to make good decisions much further down the chain of command. We're going to teach them how to gather data, how to do a quick financial analysis and know for themselves whether an idea makes sense or not. For now, we are going to have to overcome whatever is impeding them from suggesting improvements. Agreed?" asked Frank.

There were thoughtful nods around the table.

"So, job one is to treat all new ideas with respect. Next, I'd strongly encourage you to come up with a form that guides employees through the process of recording an idea, advising them about who can help them with determining the cost of implementing it and who can assist them analyzing the return on their idea. Finally, you'll want to make the path for submitting an idea very straightforward. Thoughts?" Frank asked.

No one responded. "OK," Frank proceeded, "let me ask some questions. Who makes the best forms at this site?"

There was some quiet chatter among some of the Lean Council, then Philip said, "Conchita Hispanoza. She works for me."

"Good," Frank encouraged them. "Now, who is the best one at determining the cost of a new process or piece of equipment?"

Kanisha thought for a second and said, "Jimmy Patterson."

"Finally," Frank asked, "who is the best at determining what the income implications of a decision are?"

This time it was Cheryl who responded. "That would be Tony Pearson."

"Are all of these people easy to approach and good natured?" queried Frank.

"Ah," Cheryl looked at Kanisha, "Jimmy can be a bit rough around the edges. He's not mean. He is just very straitlaced. It's easy to assume that he's being condescending."

Kanisha nodded.

"Is he coachable?" asked Frank.

"He is," responded Kanisha.

"Then let's try this," Frank offered. "Let's put the three of them on a team to develop a form that will be used in submitting suggestions. Once submitted, the three of them will review the submittals and help those submitting them to get it right before it goes anywhere else. One more thing," Frank offered, "is there someone in your organization, Cheryl, who can help this group keep the process simple?"

"Hmm," she responded, "let me ask Kanisha and Manny. Is there anyone in my organization who you think would fit the bill?"

"Ted Wright," said Manny after a second.

"Lisa Phips," said Kanisha.

"Oh yeah," said Manny. "I agree. She's a much better choice."

"Great!" said Frank. "And is manufacturing the only place where we expect good ideas to come from?"

"Funny," responded Tim, "I was thinking the same thing. It seems to me that folks in nonmanufacturing jobs can offer just as good suggestions as those in manufacturing."

"So," pressed Frank, "should we have someone on the team who can offer a nonmanufacturing perspective?"

"Absolutely," responded Manny.

"I agree," commented Kanisha. "Francine? How about you? You haven't said much."

"No, Kanisha, I agree with you and Manny. I'd like to see someone on the team representing the rest of the disciplines. Chris?"

"I'm with you guys," said Chris, blushing. He had barely spoken all day.

"Anyone else?" asked Frank.

There were no further recommendations.

"What are the last two things we always do whenever we develop a new team?"

"Appoint a leader," Kanisha offered.

"And the other?" asked Frank. When no one responded, he said, "We give them a deadline by which they need to get back to the Lean Council, right? What's that going to be?"

Everyone looked to Don. "I think two weeks should be plenty of time," he offered. "Any problem with that?"

No one spoke. "All right, let's do it this way." Don then went around the table and asked everyone individually if the team composition was acceptable and if two weeks would be sufficient. All agreed.

Don turned to Jim. "Anything else?" he asked. "Just one," Jim responded. "Could you poll the council and get feedback on the day?"

"Absolutely," responded Don. He went around the table one more time, now asking each council member what they had gained from the day, and what had they liked most. There were many variations on the theme that they had no idea how important knowing employees was to the success of an organization. Kanisha offered, "I mean I've learned a great deal just knowing my colleagues and what motivates them. I can only imagine what it will be like to understand the same information about all our employees."

When it came Chris's turn he said, "It now makes sense why no one has ever offered suggestions. I had often wondered about that. I mean, where I worked before, people were always making recommendations."

Cheryl chimed in, "True, but we're not after suggestions, we're after implemented suggestions, right?"

"Agreed," Chris responded. "Still, you can't implement what you never get. I find that really significant. I'm excited to see the new process get under way."

There were other comments, but all were in the same vein. When no one had anything else to offer, Frank turned to Jim and Don and asked, "Do either of you have anything else to add?"

Neither did and Frank released the council.

14

Tightening and Straightening

In the months that followed, Don found himself spending a lot of time with his staff. Not in meetings, mind you, but observing them and their problem-solving techniques. He sat in on their staff meetings and listened. He tried hard not to offer answers or solutions, but to ask questions.

Increasingly, his people became less autocratic and more open to listening to their own employees' ideas. When people were asked to sort out their own problems, it frequently led to self-discovery and greater self-confidence. His staff became stronger and so did their direct reports.

The other thing that made a huge difference was the introduction of top-down metrics. Frank, Jim's consultant, had come as promised. He spent almost four grueling days with Don and his staff, developing top-level metrics, identifying how they'd be measured and who would be accountable for the results.

When his own Lean council had mastered this task, they began cascading tailored versions of the top-level metrics down to their direct reports. Goals were established for each metric and direct subordinates were held accountable for their results. This same process—cascading metrics down to the next lower management team—was duplicated again and again until everyone in the organization could understand what the measures of their success were and how those rolled up to the success of the Charleston plant.

As directors began briefing from their key performance indicator (KPI) boards their organizational problems became more evident. With the discovery of their problems came a more focused approach to resolving them. The outcome was amazing. Month after month, their performance improved, even if only slightly.

KPI boards didn't just work at the director level. Every leader, down to the process leads, began using them, even in those performing office tasks: accounting, HR, engineering. As they did, they became more self-directed.

They could see their problems and could often move to correct them without intervention or help from above.

Something that Frank had taught Don was that employees rarely create most problems. "Most problems," Frank had explained, "are systemic in nature. That means that there is a system at fault. Let me explain," he went on. "You may have order entry personnel who consistently make the same errors. You may even find that some make more than others. It would seem intuitive that the ones making the most mistakes are your worst people, but that isn't a given."

"Step away from the problem and look at the system that order entry personnel are asked to use. More often than not, the system is cumbersome and overly complicated. Your people are having to work around the system and, as they do, they make errors."

"Now, ask yourself, 'What if they were allowed to redesign that system around the way the process actually works? What if their leaders told them the results that were desired and gave them an hour each week to meet and develop the new system. Even better, what if their leaders challenged them to develop Poka Yokes, mistake-proofing techniques, right into the new system.' Do you think things would improve?"

Both men knew Frank's question was rhetorical. He pressed on. "And what if you made one of them the owner of the new system? What if you held her accountable for training new personnel and being the gate through which any corrections or improvements to the system had to pass?"

Don could see his point immediately. It was as if a cloud had been lifted and he could see with perfect clarity the fact that his organization had made a practice of hiring people for their hands and essentially telling them to hang their brains at the door. "How wasteful," he thought.

"This process of changing whole systems, while definitely changing things for the better—the literal translation of Kaizen—is called *Kaikaku* (Ky-Kah-Koo). *Kaizen* is about rapid, but incremental change. Kaikaku, on the other hand, leads to radical change. Kaizen is focused over a very narrow part of a process, whereas Kaikaku takes a wider view, working on whole systems. Because of this larger scope, Kaikaku usually takes longer and involves people from multiple disciplines."

"So," Frank continued, "the Lean council might identify the order entry system as needing Kaikaku. They would handpick a leader and make recommendations to the leader about what disciplines and skill sets should be on the team. It is not unusual for the Lean council to actually name the other participants and then give the leader the ability to *voir dire* or challenge the council's picks."

It all made sense. Like everything else he was learning about Lean, the pieces all fit together like a giant jigsaw puzzle. They were, as Frank liked to say, "holistic." They grew naturally out of the way things and people worked.

Don had expected Cheryl's manufacturing personnel to be the first to use this new opportunity, but was surprised when a group of cost accountants approached the controller to ask for permission to redesign the accounts receivable process. Within two months, the number of errors dropped substantially and the average age of accounts dropped by 40%.

Not long thereafter, manufacturing did undertake a Kaikaku. Joining forces with several process engineers, a group of assemblers began working on a new assembly process that was more conducive to the tools they had and the order in which things actually went together. Their efforts took several months, but at the end, defects dropped and throughput rose. It was not uncommon for products to be assembled as much as 9% faster. Moreover, defects dropped by more than 34%. Both conditions led to happier customers and employees. Don was amazed at how much morale in the departments improved as process owners began to improve their processes. They were no longer owners in theory, but in the very way they actually conducted their jobs.

Of course, folks weren't allowed just to go change processes willy-nilly. There was a type of Kaizen called standard work that involved the documentation of the existing process, including "before" data; brainstorming improvements, implementing those improvements that made sense, gathering "after" data to ensure that actual progress had been made, creating a written/pictorial document to capture the new process, training all the stakeholders of the process, and assigning someone as a single point of accountability who would own that standard work from that time on.

As news spread of teams becoming more self-directed, more groups sought the opportunity to work on their own processes. The change was truly inspirational. Of course, the Lean council had final say over who would be allowed to do what. There were a couple of reasons for that. The first was that the entire organization couldn't be in flux at any one time. The other was that most things that needed to happen required Kaizen, and as their value stream map clearly indicated, there was an order in which processes needed to be Kaizened. Still, if the council stopped an event, those who had asked for it were informed why and given a sense of when it might happen.

Kaikakus, on the other hand, involved whole systems and all the systems were owned by the members of the council. Although they had the best interests of Friedman Charleston in mind, Kaikakus weren't free and the council had to weigh the cost versus the benefits of a Kaikaku, before they would contemplate spending time and money on an undertaking.

Tim, Don's director of continuous improvement, was responsible for five things:

- Training the entire organization—including Don and his staff—in Lean philosophies and methodologies
- Creating and maintaining the plant's value stream map and identifying opportunities for improvement to be voted on at the Lean council
- Conducting Kaizen events intended to gain quick increases in safety, quality, throughput, and cost
- Conducting Kaikaku events intended to design and revise whole systems, generally spanning multiple organizations
- Working with leaders throughout the organization to monitor, sustain, and improve gains made during Lean events

Tim, as did all CI directors, reported directly to the plant manager, in this case, Don. In addition to their own half hour together each week, Tim was a voting member of Don's Lean council and accorded equal standing with the other directors.

For the first six months, Tim conducted an average of two Kaizen events a month. After a few months, Don realized that was a burnout pace and authorized Tim to hire a CI coordinator who would conduct at least half of the events. In time, Don would allow Tim to invite at least two employees at a time to become six-month interns. The interns did a lot to help him and his coordinator prepare for events, run errands during events, and over time, perform training.

The other cool thing, Tim quickly realized, was that the interns became a bit of subject matter experts, so that when they returned to their departments, they often jumped the department's Lean status to a whole new level.

Tim also sat in on every Kaikaku meeting and helped teams to navigate the political shoals of working on systems that no one personally owned.

After the initial Hoshin Kanri, Frank conducted what he called maintenance visits. He would show up once a month for a day or two, take notes on all the KPI boards, measure progress, and work with Tim. Each visit also included a private discussion with Don and a joint phone call to Jim.

Don's direct reports were also scheduled to visit other plants to understudy their counterparts there. After those visits staff members returned with static electricity crackling all around them. They would hold training sessions with their people and soon the entire discipline would be developing lists of new behaviors and best practices.

Don admitted that those events were expensive, but his controller kept records of all the improvements made after such visits and there was a pretty consistent 3:1 return on the initial investment. Actually, at Jim's recommendation, Don had his controller treat the CI group as its own profit center. They measured event cost versus new revenue and cost reductions. There was a substantial positive cash flow for the group.

15

X-Matrix

A month before Frank's next scheduled visit, Don got an e-mail from him. "There's a second component of Hoshin Kanri," the e-mail began. "We have reviewed and implemented the component that deals with cascading metrics. The second component involves creating a long-term breakthrough strategy, a true north, if you will."

The note had gone on to say that Frank was going to need Don and his staff for as much as a full week. He gave Don the date of his arrival and asked if they could reserve the conference room. Don knew this wasn't really a request. Frank spoke with Jim's full authority. Don dutifully notified his staff and asked them to keep their calendars clear that week.

This wasn't Don's first rodeo involving strategic plans. Like most executives, he had developed them in the past. He would do this because it was expected of him, but if it was anything like what they'd done in the past, the plan would end up on a bookshelf until they updated it a year from now. He'd try to keep an open mind, but he wasn't looking forward to the exercise.

Four weeks later, when everyone had taken his or her seat, Frank began. "How many of you know what we're going to be doing for the next week?"

Jennifer, Don's sales and marketing director, raised her hand. "We're going to create a strategic plan," she stated.

"And what does that mean to you?" Frank asked.

There was silence. Finally Cheryl stated what was on everyone else's mind. "We're going to waste a week."

"I wish you would be more direct," Frank said with a chuckle. "You're right, though," he confessed. "For many of you, a strategic plan is an exercise, something to be able to point to and say 'Yeah, we did that.' OK, let me ask you another question," Frank started again. "Have I ever wasted your time? Do I impress you as someone who would waste your time, and by implication, my own?"

Again, silence. Don grudgingly stated, "No, you haven't and you don't impress me as someone who would. It's just that we have some bad history to overcome."

"Great!" said Frank. "Honesty. I couldn't ask for better than that. Let me make a commitment to you," he continued, looking at Cheryl before the others. "This will not be a waste. What we are about to do will be used every day in the years ahead. Its use will become second nature to you. Knowing that, will you give it a try?"

There was murmured assent around the table.

"OK," Frank picked up, "let's get going." He walked to the whiteboard and drew a huge 'X.' "This," he announced, "is the beginning of what Toyota calls an X-matrix. It is part of an organization's Hoshin Kanri, and when we finish, it will be the master plan for the next three to five years of your business." Frank returned to the whiteboard and completed the framework of the X-matrix.

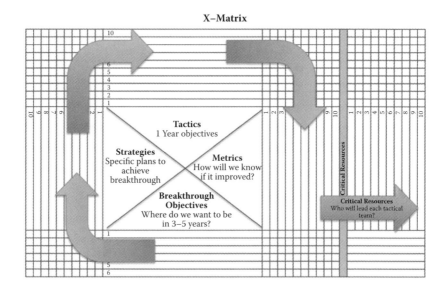

X–Matrix

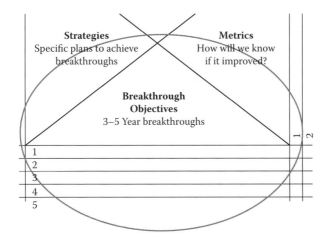

"So this is the framework we're going to use to document your strategy. Doesn't seem all that imposing, does it?" There were headshakes around the table. "That's the easy part. Here's the tough part. You actually have to identify your breakthrough objectives, develop a strategy to achieve each, and then break each strategy down into a series of tactics that will lead to its accomplishment. You'll also have to determine how you'll measure your successful accomplishment of each tactic. Finally, you'll need to identify, by name, the person that you're going to hold accountable for achieving the tactic."

"Like everything else in Lean, completing an X-matrix flows logically, so we just need to get started. Let's begin with where you want this plant to be in five years. Any thoughts?"

"Actually, I've given this some thought since your e-mail," Don stated. "We're seeing a growing number of designs calling for plastic cabinet construction. Because of weight and cost, and the fact that plastic is non-conductive, industry analysts are forecasting that will be the direction in which the industry will continue to move. We've already seen an almost complete shift to plastic in residential construction, so the direction seems logical. To go down that path, we'll need to purchase the equipment to compete. We'll also need to hire or train people who can run and troubleshoot that equipment."

Jennifer chimed in, "Several of our competitors have added injection molding equipment. The size cabinets we're seeing are shrinking, too. They are incorporating more technology, but it's shrinking at the same time. Fortunately, our competition's quality can't compete with ours, but it's clear that the market is moving in that direction."

"Anyone else?" Frank asked. No one responded. "OK," he continued, "we're a long way from being finished, but let's start here and see how this X-matrix works." He went to the board again and below the section titled "Breakthrough Objectives," he wrote "Enter Plastic Cabinet Market." "Anyone else?" Frank asked.

Again, Don spoke up. "We have got to drop our manufacturing through-put time. It takes us too long to make a system. That affects our cash flow, because the time from the release of material to the shipping of the final product takes so long. We've got to get a handle on it."

Frank wrote on the whiteboard and turned.

After a brief silence, Philip raised his hand. Frank nodded at him and Philip began, "Our employee turnover is almost 30%. That's after we factor out all the changes associated with Don's new staff. We've got to do something to make Friedman an employer of choice in the Charleston market."

Several in the room applauded Philip for exposing what had clearly been a concern for many.

"Very good," Frank praised the group. "Any others?" When none were offered, he said, "Now, what strategies do you need to undertake in order to meet those breakthrough objectives?"

There was silence for a second, then Jennifer offered, "We'd need to purchase an injection molding machine."

"Are you sure that's the piece of equipment you'll need? Why not a compression molding, or a blow molding machine?" Frank asked.

There were looks exchanged around the table as if to say, "Gee, I don't know."

"You know I don't let silence speak, so let me ask you this. Doesn't it make sense that one of your strategies would be to study the type and size of plastic cabinet to make?"

There were nods of agreement.

"How long do you think it will take to conduct that study?" Frank asked.

Kanisha spoke up. "At least six months," she said.

"So, we'll have to wait at least six months before we can even purchase the equipment?" Jennifer asked. "When do you expect to actually make plastic cabinets?" Her question came out as accusatory, as if Kanisha was going to be holding the plant back.

"Wait," Kanisha shot back, "you're just telling me now that our competition is using plastic cabinets. How was I supposed to know? Have you done any market studies of our competitor's cabinets? Are they any good? How is the market responding to them? What equipment are they using?"

"Do you really expect me to just go pick out a machine on your say so?" Kanisha was getting hot. "Do you have any idea how much we could be talking? It's well over $1 million. Do we have $1 million in our capital plan?"

"All good questions," Frank intervened, giving the parties a chance to cool down. "I think what Kanisha is trying to suggest is that we need to do some due diligence before we are even ready to specify a machine, much less order it. Might those be our year one strategies? If so, what do we need to know?"

Kanisha started again, "For sure, we need to know if plastic cabinets are even being well received in the market. Another thing, do we intend to mold an entire cabinet, or are we going to mold components and assemble them into a cabinet? That will determine the size, and possibly, the type of the machine."

Turning to Francine, she asked, "Do we know what plastic compound our competition is using?" Francine did not.

Frank went to the board and wrote "Perform Market Analysis" in line one of the strategy blocks. Off to the side he wrote, "Determine Cabinet Details (composition, size, assembly)."

"Anything else?" he asked.

"When we know those things," Kanisha began, "we'll need to spec the machine and determine utility requirements. We may need to purchase a larger electrical distribution service and possibly a new air compressor. We'll need to identify plastic suppliers and hire someone who will design the molds."

"Oh, and will we be exhausting new volatile organic compounds (VOCs) into the atmosphere?" she asked. "If so, we'll need to submit an emissions permit request to the state and hope we can get permitted. That should be done before we order the machine." She stopped. "That alone is a major undertaking," she told the team. "Those studies alone can take a year. Are we sure that the market niche is worth all that?"

Jennifer started to answer, but Don cut her off. "Industry forecasts show that plastic cabinets will comprise between 30% and 50% of the market in 10 years. We need to start positioning ourselves to be ahead of the competition. What's out there now are just small boxes, not whole cabinets. Forecasts show the market shifting toward plastic cabinets for a number of reasons." Jennifer nodded her assent.

Frank jumped in. "So, year one needs to be a year of study and analysis, right?"

"Yes," Don replied.

"OK, that's good. Because you'll only be doing analysis this year, will you need to alter any processes?"

"I don't think so," replied Don.

"OK, then, before filling in all the tactics, let's look at what other strategies you'll need to undertake. Anyone?"

"I'd like to start qualifying suppliers," Francine began.

"Tell us what you mean," said Frank.

"Well, right now we have some 3,500 suppliers in our database. I'd like to reduce that to under 1,000. I'd like those who are left to have to demonstrate their ability and willingness to meet our quality and delivery needs. I want to switch our procedure from buying from the cheapest supplier, to purchasing the product with the lowest cost of ownership. That will mean measuring every supplier's on-time delivery and their ongoing quality performance. It will mean that we send engineers in to inspect their facilities and processes. We'll need to set aside buyer time to study supplier statistics and begin a more proactive relationship with them. I'm seeing that as being a year or even two years to complete."

Frank stopped her. "Do you see any of these being required to give you a strategic market breakthrough?" he asked.

She cast her eyes down. "No, I guess not."

"Please," Frank encouraged her, "don't let that impede you from pursuing that as a major initiative. All I'm saying is that it doesn't pass muster as a breakthrough objective." He stopped. "Don, Tim, Kanisha, Francine, do any of you see this altering your market position?"

All agreed that it did not.

"Frank," Don began, "I wouldn't have come to this conclusion on my own, but I can see the wisdom in what Francine is suggesting. I don't want to leave here without her knowing that I would like her to pursue this initiative."

Francine smiled her appreciation.

Manny, the controller, spoke up. "Corporate is moving all facilities to a new database. It will integrate all disciplines, be real-time, and provide much better search functionality. That's going to be a sizable undertaking and will need at least one point of contact (POC) from each discipline. Corporate has us scheduled to go live next year."

Again Frank asked, "Manny, same question as I asked Francine. Will the implementation of this new system alter your market position?"

"Not really," Manny replied.

"Then, for the same reason, it doesn't belong on this list, but you'll still have to do it," he said with a smile. "Anyone else?"

Cheryl had been silent, but now offered, "We really need a new paint system. We're still using wet booths and the EPA is constantly on us for exceeding our VOC limits. Plus, because we're using cascading waterfall technology, we have to deal with the disposal of the overspray as hazardous waste. That's expensive and time consuming."

"Good call," said Kanisha. "I'm glad you remembered."

"Breakthrough worthy?" Frank asked.

"Actually," Don offered, "it is. Paint is one of the longest elements in the entire manufacturing process. This actually complements the breakthrough objective of reduce throughput time. I believe it qualifies as a strategy for achieving that objective."

"Very good," Frank praised him. "I'd have to agree."

Frank had been writing down each breakthrough objective and strategy on the board. "Last call," he announced. He went around the table and asked each person if he or she had any other contributions. No one did.

"OK," Frank continued, "three breakthrough objectives is just about perfect. Now, if you could only do one, which would you choose?"

There was debate around the table. Plastics was their future, and they would have to pursue it if they wanted to compete, but paint was killing them now.

Philip made an argument for employer of choice. "We have a really bad reputation in the Charleston market. We pay a competitive wage, but have difficulty attracting the best candidates, especially for technical skills. That affects us today, too."

There was healthy debate, but they ultimately chose the paint booth. Paint was involved in their day-to-day operations and could not be put off. "Besides," Cheryl stated, "if we voluntarily reduce our VOCs now, the State EPA may be easier to talk with when we approach them about our new plastics processes."

"Good point," Don agreed.

Frank asked, "If you could do only one more, what would it be?"

More debate. Finally it was agreed that plastic cabinets were critical to their future and that they had to get moving on that.

"And only one more?" Frank asked.

That left employer of choice as third. Before moving on, Frank said to Philip, "There's no shame in being third. It will still be pursued. What we really just decided is the order in which we choose critical resources later in the process. Essentially, we have agreed that we'll allocate the best project leaders to the paint system and plastic cabinet objectives."

"All right," Frank summarized, "here's what you just agreed to."

BREAKTHROUGH OBJECTIVES

1. Reduce throughput time (paint system)
2. Plastic cabinet
3. Employer of choice

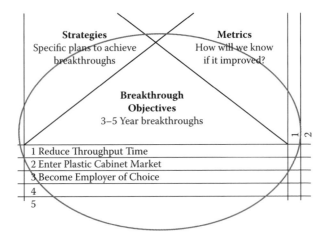

"If there are conflicts with resources, you'll resolve them using this priority list. Can you stick to that?" Frank asked Don.

"As long as corporate doesn't renegotiate them for us," Don said. "Not that that would happen," he said with an impish smile.

That comment stopped Frank. He looked at Don and said, "As soon as we finish this process, you need to get on the phone and notify Jim of your intentions. Explain your rationale and solicit his support. You're right," he concluded, "corporate support is critical and Jim is your best path to getting it."

Jennifer raised her hand. "Could we have a break?" she asked, looking at the clock. It was already 10:30.

"Sorry," he said. "As you can see, I get all wrapped up in this. Now that you mention it, I could use a break, too," he admitted. "Fifteen minutes? That puts us back in here by 10:46."

When the team had reconvened, Frank began. "Next step: strategies. What strategies do you need to engage in so as to achieve your breakthrough objectives? Keep in mind," he challenged them, "whatever you decide needs to be accomplished in the next 12 months. One year. That's all you get."

Cheryl and Kanisha had made such a compelling case for the new paint system they all agreed that it really needed to be their top strategy for reducing throughput time.

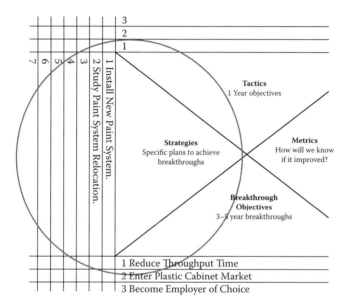

Cheryl then surprised the group. "Kanisha," she asked, "do you have any sense of footprint for the new equipment? Will it be generally smaller or larger?"

"No idea at this point. Why?"

Cheryl addressed the entire team. "The paint system is a real monument, big, heavy, and immovable, and we'll be tinkering with it. Right now, its location is illogical given where it falls in the value stream. Do we want to consider whether it would make sense to install the new one somewhere else in the building? I mean, its current location makes us backtrack after mechanical assembly. That adds a lot of movement waste, and we make up for it by having a lot of WIP in front of paint, so they don't run out. Should we use this opportunity to change its location in the plant?"

"Good point," said Don, "but does it require its own strategy?"

There was a brief discussion, but Cheryl made a cogent argument for the two requiring two totally different viewpoints. "You've got one team focused on spec'ing and buying the new system," she said. "Why burden them with deciding where to locate it? Kanisha?"

It was becoming evident that these two had bonded. Kanisha said, "I think you're right. If the team spec'ing the new equipment can provide you with utility and building requirements, I see no reason why you can't handle these separately. I will say, however, that we'll need the recommendations of both before we go after capital."

In the end, the team agreed to handle them separately.

"Oh," Frank stopped the discussion, "from this point on, I'm going to take you through the rest of the X-matrix using only your number one strategy. We'll do that one completely, so you see how the X-matrix works, then we'll come back and complete the rest."

"Are we ready to move to tactics?" Frank asked. "Again," he admonished them, "tactics need to be accomplished this year."

There was another spirited discussion and the team finally agreed on "identify new painting system, supplier, and installer" as their primary tactic. There had been some discussion about making "install new paint system" the tactic, but Don reminded them that they'd need to submit a request for capital and have it approved first. The team agreed that it would be a stretch to get all that done and install the system in a single year.

"Frank," Don began, "we really can't define the capital needed for the new paint system if we also plan to relocate it. Doesn't it make more sense to treat the relocation of the new system as a separate tactic of the same strategy?"

"Your call," said Frank, "but it's pretty important and could alter the amount of capital you go after. Do you want to create a separate tactic and team to study how you might realign your entire manufacturing process?"

More debate followed, with one group arguing that it should be separate and a second one arguing that the same team should study both. Finally Cheryl said, "I see totally different skills being needed by the two teams. I vote that we keep them separate."

"Agreed," chimed in Kanisha.

"Any dissent?" asked Don.

No one spoke.

"What's the rule about silence?" Don asked, cutting Frank off with a raised hand a là Jim. The other man smiled broadly.

"It can't speak for us," mumbled Manny.

"So?" asked Don.

"Separate," said Manny.

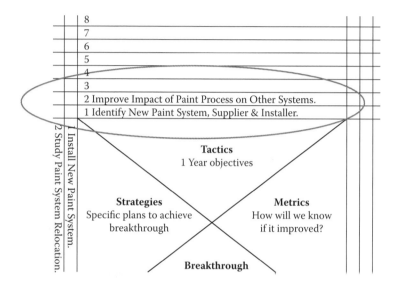

"Separate," said Jennifer and Francine. Don finished by going around the room to complete the vote. "Separate," he announced at the end of the vote.

"Separate it is," said Frank. "How do you want the tactic to read?"

This discussion didn't take long. Everyone pretty much agreed on "improve the impact of the paint process on other systems." That, they felt, was broad enough to allow the new system simply to replace the old one in the same location, or relocate it all together.

"Is that it for your first strategy, then?" Frank asked.

There were head nods around the table.

"Before moving on, let's look at these little boxes formed by the intersection of the breakthrough events and strategies rows, and again at the intersections of strategies and tactics rows. We use these to indicate which strategies support which breakthrough opportunities and which tactics support which strategies. So let's go back to put Xs in the appropriate places."

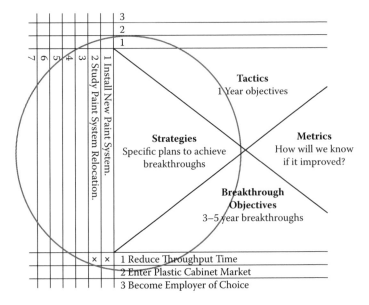

"Does 'install new paint system' support any other breakthrough event than 'reduce throughput time'?" Frank asked.

The team agreed that it did not, and Frank placed an X at the intersection of the row and column. Frank pressed on, this time asking, "Does 'study paint system relocation' support any other strategy?"

Again, the team agreed it did not, and Frank placed an X in the box formed at the intersection of the two. Frank continued through the intersections of the tactics and strategies rows, placing Xs at the intersection of each. "Anything else?" he asked.

There were no other suggestions. Frank looked at his watch. "Lunch should be here soon. Take a quick break."

16

Metrics and Critical Resources

When everyone returned from lunch Frank slapped his hands together and said, "All right, let's get back to the discussion of what needs to change in the paint process and how we'll know if things are getting better. In doing so, I want you to set some really aggressive goals for the folks spec'ing the new paint system. How will you know if you improve things? What would you measure that would tell you?"

Cheryl had been waiting for this question. "We'll measure the key performance indicators (KPI)," she blurted.

"Good," Frank complimented her. "And they are…"

Cheryl was unprepared for the question. She shrugged her shoulders.

"Tim," Frank asked, "do you have a copy of the value stream map for the entire factory?"

"I can get to it quick enough on the shared drive," came Tim's reply. He opened the file on Frank's laptop and projected it on the screen.

"What are these spiky balloon-looking things above the paint system?" Frank asked.

"Opportunities to improve the future state," said Don with renewed understanding.

"What kinds of things did you think you needed to improve, Cheryl?" Frank asked, although the answer was on the screen for all to see.

"It's right there," said Cheryl.

"Humor me," replied Frank.

"It says that changeover time is 4.5 hours, that cycle time is 3 hours and 20 minutes, that downtime is 15% of available time and that defects are 10%," Cheryl responded.

"So," Frank asked, "what kinds of things do you need to improve?"

"Got it," said Manny, as if a light had come on. "So," Manny continued, "at a minimum, we need to improve the following attributes of the paint process:

- Defects
- Cycle time (CT)
- Changeover time
- Downtime

"Let's also look at the triangle in front of the paint process," directed Frank, pointing to the value stream map. "What does it suggest?"

"That we have 15 hours of work in process (WIP) in front of paint," Don responded.

"Walk me through the logic of why all that WIP is there, Cheryl," Frank requested.

"Why me?" she asked with a face suggesting that she was innocent of whatever Frank was driving at. It drew the expected laugh from the assembly. "OK," she began, feigning reluctance. "WIP is a reflection of the length of changeover time. The more WIP, the longer the changeover time. If we consult the VSM, changeover time is long, so it is logical that WIP would also be high. And, of course," she said pointing to the WIP triangle, "it is."

"OK," Frank took back over, "if you have 15 hours of WIP in front of paint, how many parts do you have in WIP?" He was met by blank stares. He went to the board and wrote:

$$15 \text{ hours} \times 3{,}600 \text{ seconds per hour} = 54{,}000 \text{ seconds}$$

"What's the cycle time of the paint process?" he asked.

"3 hours and 20 minutes," replied Cheryl.

"Good, but do you only hang one part every 3 hours and 20 minutes?" asked Frank.

"Of course not," came Cheryl's response.

"OK," Frank asked, "what goes on during the cycle time, then?"

"Isn't cycle time comprised of the time it takes to hang the part, paint the part, cure the part and remove the part from the overhead rail. Right?" she asked.

"Right," said Frank. He drew and labeled four boxes on the board.

Hang	Paint	Cure	Remove
42 Sec	3 Min	3 Hours 15 Min	42 Sec

"How long does it take to hang the part?" Frank asked.

"On average, it takes about 42 seconds," Cheryl responded. Frank added the value below the operation name.

"How long does it take to paint a part?" Frank asked.

"On average, about 3 minutes."

"How long does it take to run the parts through the curing oven?" Frank asked.

"Almost 3 hours and 15 minutes," came the response.

"And to remove the part?" Frank asked.

"42 seconds."

"So," Frank asked, "is it safe to say that every 3 minutes and 42 seconds—the time it takes to hang and paint the part—we're taking a part out of the incoming material?"

"Yes," Cheryl agreed.

"Do the math," Frank instructed. "You have 54,000 seconds of parts in queue. If you process them at the pace at which you hang and paint parts, how many parts does that mean you have in queue?"

This time Manny had his calculator out and was rapidly tapping the keys. "243 and change," he said.

Frank handed Manny the marker and motioned him to the board. Manny wrote:

$$3 \text{ min} \times 60 \text{ sec/min} = 180$$

$$180 \text{ sec} + 42 \text{ sec} = 222 \text{ sec/part}$$

$$54,000 \text{ sec}/222 \text{ sec/part} = 243.24 \text{ parts}$$

He looked at Frank who gave him a nod. Manny gave Frank back the marker and returned to his seat.

"Now," Frank asked, "do you really only paint one part at a time? Most places I've been daisy chain the hangers, hanging one on another on another so they may have as many as six or seven parts all daisy chained off one another. Any chance you do that?"

Cheryl admitted that they did.

"What's the average number of parts in a daisy chain?" Frank asked.

Cheryl thought for a second. "About five."

"So, back to Manny's calculation, we need to multiply his answer by five, don't we?" Getting head nods he continued. "So,

$$54,000/222 = 243.24.\ 243.24 \times 5 = 1,216 \text{ and change}$$

You have over 1,200 parts sitting in front of paint at any one time. Can we agree that you have a lot of opportunity for improvement?"

There were wry smiles and head nods around the table.

"Let's think ahead. If we reduce setup time, we have the capability to make more parts than our customers have ordered, but that doesn't mean that we should make them. If we didn't, we'd have excess labor capacity. What could we do with that?"

"Build ahead and stop early," Manny offered.

"Build to Takt time, and have the employees 5S their area," suggested Francine.

"Use the employees to do something else that adds value," replied Tim.

"Ah, very good," Frank said pointing to Tim. "Give me an example."

"Well, we could rebalance the line so that we could remove a person at least part of the work day."

"Do we ever make more parts than the customer has ordered, Manny?" Frank asked.

Manny was quiet for a second. "In the pre-Lean world, I'd have said 'Yes,' but I know that the Lean answer is 'No.'"

"Can you live with that?" Frank asked.

"I'll be honest," Manny responded. "It translates to intentionally giving up that capacity forever. What happens if something in the system fails and the paint line goes down? It makes me twitch … " The room erupted in laughter. "… But," Manny continued, "the logic is sound for why we wouldn't. I just need to look at the manufacturing process through new lenses."

"Well said," Frank praised the controller. "OK, back to how we'll know if we're improving. What should we measure?"

Kanisha ticked off the list they'd just developed:

- Defects
- Cycle time (CT)
- Changeover time
- Downtime
- WIP

"You know me," Frank stated, "I'm lazy and don't want to do anything more than I have to. Are any of these lagging indicators?" He was met with blank stares. "Manny, remind us what leading and lagging indicators mean."

"A *leading indicator*," the controller began, "is one that is not a derivative of anything else. A *lagging indicator* actually follows another indicator. First the leading occurs, then the lagging follows."

"Great explanation, Manny. So, are any of the KPI above derivatives of any others?"

"For sure WIP is a derivative of changeover time (C/O)," he replied.

"Good," Frank praised him again. "Any others?"

"Not really," replied Manny without hesitation.

"OK," Frank stated, "then we need to measure four. Sounds like a lot, but I'm going to show you how to do it easily. Before moving on, let me ask: is paint the only process that will need to change? Francine, will a change in the paint system change anything in the procurement process? Will it change anything in the receiving or quality processes? Kanisha, will it change the EHS processes, or the engineering process?"

"As I said earlier, a new paint system will absolutely require us to submit a new emissions permit request to the State EPA." Turning to the others she continued, "I can see Frank's point. Changing the paint system from wet to dry will have ripple effects on other parts of the company."

Frank picked up, "I'm not sure that they all need to be tactical changes, but changing the paint system will definitely have an impact on the rest of the organization; and it will steer the selection of people for inclusion on the assessment team. We know what we'll need to measure. Let's set some goals. Tim, do you want to start us off?"

"Well, we'll want to reduce all four metrics: defects, cycle time, changeover time, and downtime."

"OK," Frank began, "what would be an aggressive goal for the paint system tactical team?"

"Shouldn't we know what the equipment capability is before we set that?" asked Kanisha.

"Not really," replied Frank. "If you set the goal based on the equipment capability, you'll only get what is currently available in the market. Will that propel you ahead of your competition?"

"OK," she said, squinting, "but I'm keeping an eye on you." That drew laughter.

"Goal for defects?" Frank asked again.

"5%," Kanisha said. "That's a 50% reduction."

"You can do that with your eyes closed with a new system," shot back Frank. "I'm keeping an eye on you, too," he said smiling.

"3%?" offered Cheryl.

"Great suggestion, Cheryl," Frank shot back. "1.5% should be just aggressive enough."

Cheryl's jaw dropped. "Wait a minute…" she started, but she knew she'd been set up.

"Cycle time?" Frank stated. "What do you think keeps the CT so high?"

"Easy," responded Cheryl, "it's the old math problem about a train that leaves the station at 1:00 o'clock and travels 50 miles an hour; how long before it arrives at the station 80 miles away? The conveyor is a fixed length and moves through the oven at one speed. The speed is based on the time it takes to bake the paint without causing blisters or burning."

"Is a conveyor oven the best way to cure the paint?" Frank asked. "I mean, are you slaved to use a conveyor oven, or can you use wheeled racks and push them into small ovens for a fixed amount of time?"

"I'm not suggesting that you take that approach," Frank assured them, "although I've seen it used effectively in small volume paint systems. What I am suggesting is that you set a CT ambitious enough that it makes the team and the suppliers think beyond what's easy. In short, just because that's the way it's always been done, don't believe that it's the only way to do it in the future. Someone want to suggest a CT for the new system?" he asked.

"How about an hour?" asked Kanisha.

"This time, I agree with Frank," said Don. "I think we can do better. I'd say we ought to shoot for 30 minutes."

"Anyone else?" Frank asked. When no one made a suggestion, he said, "That's the best you think you can do? OK, you don't get out much, but I see lots of plants. I'd recommend you shoot for cycle time ≤15 minutes."

"Easy for you to say," mumbled Cheryl. "You won't have to do it."

"Fair enough," responded Frank. "I'll tell you what. I'll start you out with the name of a paint system manufacturer that will get you in that ballpark. Deal?"

Cheryl brightened visibly, but she still thought it was too aggressive.

"Changeover time?" Frank asked next.

"Fifteen minutes," Don threw out.

"That's pretty aggressive," Frank acknowledged. "Can anyone tell me why changeover takes so long now?"

Cheryl began, "Well, they mix the paint in large vats and park them outside the paint booth. Every time they change colors, the lines need to be flushed and the tanks and guns cleaned. Maintenance uses that time to clean the weirs in the waterfall system."

"OK," Frank agreed. "Will you be going with a wet painting system?"

"Probably not," she admitted. "If the new system is dry, we'll use mechanical filters."

"How about mixing paint?" Frank asked.

"Good point. If it's dry, the cleanup can probably be shortened."

"Look," Frank began, "I don't want to design the system here, but if it's dry, why would you mix it at all? Can't you get the suppliers to provide it to you already mixed?"

"I suppose so," she admitted.

"Why are you mixing so much anyway?" Frank asked.

"Because…" she stopped herself before completing the sentence, "… because the changeover takes so long, we try to *amortize* the setup over as many parts as we can."

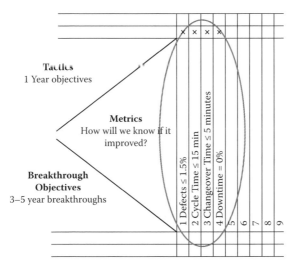

"So," Frank concluded, "how long should changeover take?"

There was a fresh debate before the team agreed on ≤5 minutes. Frank wrote that value next to the other two goals.

- Defects ≤1.5%
- CT ≤15 minutes
- Changeover time ≤5 minutes

"So, what's left? Ah, downtime," Frank proclaimed, as he examined the original list. "OK, who wants to start the discussion … Francine?" Frank asked the startled materials director.

Francine stammered for a second or two before answering. "I don't even know where to begin," she admitted.

"This is easy," Frank grinned. "Ask yourself, in a perfect world, how often should a piece of equipment be fully operational?"

"Put that way," Francine smiled back, "it is easy. It should be ready 100% of the time."

"Good," Frank praised her. "And if it's ready 100% of the time, how often would it be down?"

"Never," she responded.

Turning to the whole group Frank said, "Let me give you a scenario and ask you to fill in the blanks, so to speak. First, how many of you own cars?"

They all raised their hands. "OK, what is needed to keep a car ready to be used 24/7/365?"

"Keep gas in the tank, change the oil every 3,000 miles, rotate the tires at every oil change, and perform scheduled maintenance faithfully," said Philip.

"Excellent!" complimented Frank. "And if you do those things, what spare parts should you keep on hand?"

Philip thought for a second and said, "Spare tire."

Frank gave him a quizzical look. "You don't keep a spare battery and generator?" he asked.

"No, why would I?"

"So, Frank, here's what I think you just told me. If you perform routine preventive maintenance and keep only those parts with a high incidence of failure, you have a reasonable expectation that the car should be ready whenever you are. Is that about right?"

Philip thought again before answering. "Yes."

"OK, one last question," Frank stated. "Do you paint 24 hours a day?"

Philip looked at Cheryl who answered the question. "Rarely."

"So, is there any reason that you wouldn't have time to maintain the paint line preventatively?"

"Put that way," Cheryl replied, "we should always have time." Turning to Kanisha she asked, "Would your folks maintain it on third shift?"

"Don't see why not," she responded, "as long as they get comp time."

"So," Frank picked back up, "is it reasonable to believe that the new paint line should never go down unexpectedly?"

"It is," said Kanisha.

"In that case, what should downtime be as a percent of available time?"

"Zero," she responded.

Frank completed the list.

- Defects ≤1.5%
- CT ≤15 minutes
- Changeover time ≤5 minutes
- Downtime = 0%

Turning back to the team, he asked, "Do you see what we've done? We've created a list of clearly defined expectations for the team choosing the new paint system. Ask yourselves, 'when the system is installed, will we be able to tell whether it meets our expectations?'"

Don smiled, saying, "I really like this. We've not only defined what we need to do, but we've also created metrics that will tell us if we've achieved our objectives."

"And," Frank added, "we're not finished. "The last thing we're going to do with the X-matrix, is to define the critical resources we'll need in order to accomplish each tactic. What we're looking for here is the name of the team leader who will be responsible for delivering the results we're asking for. Occasionally, we'll need to tie up a critical piece of equipment or a person with unique skills for protracted periods of time. Even though they're not a team leader, this portion of the X-matrix is a good place to record those decisions. Before I go on, does naming someone to be accountable for the tactic make sense?"

There were head nods around the table.

"OK, then let's complete paint," Frank instructed. "Who should be the team leader responsible for delivering the new paint system?"

"Isn't that obvious?" asked Kanisha. "Cheryl."

"Is Cheryl able to devote time to contacting suppliers, holding meetings, and generating specs?" Frank asked.

"Heck NO," said Cheryl.

"My recommendation," said Frank, "is that you choose someone who is a good project manager. It would be nice if they also knew about paint systems, but it's their organizational skills and the ability to keep a project on schedule that will mean the most to you. Now, with those parameters, is there anyone else you could recommend?"

"One of my engineers, Samantha Brown," offered Kanisha. "She fits the bill. Great organizational skills, able to move projects along even under adverse conditions."

"Quint Sheppard, one of my production controllers, fits the bill, too," interjected Francine.

When no one else offered further recommendations, Frank asked each woman to explain why their candidate made sense. When both had, he called for a hand vote and Samantha was chosen.

"Halfway there," Frank pronounced. "Who should lead the study of the paint system relocation?"

"It's not that I don't have someone," Francine interjected, "it's just that I'd like him to lead the study of the plastic cabinet tactic. I'd personally be in favor of Quint heading the paint system relocation study."

"Is that it?" Frank asked. "Does anyone have any other recommendations?"

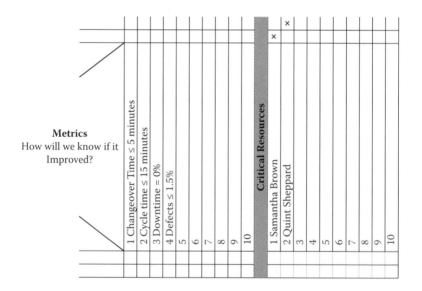

He went around the room asking, but there were no other recommendations, so Frank wrote Quint's name on the X-matrix.

"Great!" Frank proclaimed. That's enough for today. You have now taken one strategy all the way to completion. Tomorrow, with a little luck, we'll finish up, and then we'll move into talking about how these teams will report back to you. Anybody had enough?" he asked. There were groans around the table. "Eight a.m. tomorrow," Frank stated.

The team exited silently, and from the looks of them, it had been another hard, but fulfilling day.

Frank turned to Don, "Care for a drink?" he asked.

"I thought you were a teetotaler," Don remarked.

"Not after work," Frank assured him. Quoting the Apostle Paul he followed with, "All things in moderation."

"Still OK to leave things set up in the conference room?"

"Absolutely," replied Don. "You willing to let me show you a good watering hole?"

"If you insist," Frank shot back with a grin. "First round's on me."

17

A3 Report

A good night's sleep had restored the team's enthusiasm. Frank had pushed them hard the day before, but the worst was over. They now knew how the X-matrix worked and now all they needed to do was to slog through the completion of what remained. Frank knew that was still going to be a challenge, but also knew this team was up to it.

For the most part, Frank stepped aside and let Don run the show. The two men had agreed on this approach the night before over a beer. It was enjoyable to watch Don work. He was very different from Frank, but got the job done. Don had argued the night before that, if he were to update the X-matrix by himself next year, it would be nice to have a chance to do it under Frank's tutelage now. That had actually been Frank's plan, but he let the other man have the credit. By day's end, they'd completed the X-matrix through the metrics block. All that was necessary now was to identify the critical resources, or, as Cheryl had started mimicking Frank, the "belly buttons" accountable for making things happen.

Day 3 was a tough one. It wasn't hard to identify the people who would be good at leading each of the improvement teams. The haggling came in two forms: those who wanted their candidate to head a particular team, and those who didn't want to lose their direct report, even though the rest of the team deemed that person best for the role.

Frank finally had to intervene. "Remember," he enjoined them, "we're talking about the people who will determine this plant's future success. If one of your people is the best candidate, consider giving one of their subordinates an opportunity to fill those shoes. Not only will you get your job done, you'll deepen your own bench. To get them up to speed, assign someone else in your organization to mentor them."

It was 3:30 when the last resource slot was filled, and Frank sent them home.

"Had enough?" he inquired of Don.

"I'm pooped," the other man admitted, but "I'd be happy to get feedback over a beer. This time it's my treat."

The team assembled quietly the next morning. They were showing the effects of mental fatigue. Frank knew this had to be the last day, but also knew that this team had drunk deeply from the well of Hoshin Kanri. This event would be emblazoned on their minds forever.

"Last thing, I promise," he stated.

"You'll want to have one way that each of your tactics teams will report to you. Think of it as standard work for reporting. So, each of your critical resources…" he paused, "what do you call them?" he asked Cheryl.

"Belly buttons of accountability," she said with a chuckle.

"So," Frank picked back up, "each of our belly buttons of accountability will need to brief this team once a month using this standardized format. Next week, I'll expect you to conduct a training for the collected critical resource team. That means that what you learn today, you'll teach next week. We clear?" he asked.

He was met with apprehensive head nods.

"The format we'll use is one that is used around the world by Lean practitioners. It's called the A3. It gets its name from the size paper it is written on. A3 paper is 11" x 17" and is traditionally oriented in landscape mode. Here's what a blank one looks like," Frank announced, projecting one up on the screen.

"Ready to teach it?" he asked.

Several in the room proclaimed "No way!"

"So, you want me to explain a little more?" Frank asked.

"Please," said Philip.

"First things first," Frank began, "if you Google 'A3 Report' you'll get a vast number of variations. This is what is called a seven-box A3. We'll talk more about that in a bit. What do you think goes in the title block?" asked Frank with an impish glint in his eyes.

No one spoke. "Let's try this again," he said. "What is the first tactic that we'll be pursuing?"

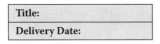

Manny rifled through his notes. "Identify new paint system, supplier, and installer," he pronounced.

"Excellent, Manny," Frank complimented the other man. "So, if you were a betting man, Manny, what would you think the title of this A3 would be?"

"Identify new paint system, supplier, and installer," he responded.

Frank handed around blank A3 forms. "Let's fill this in," he instructed. "If you have concerns that you won't remember why you did something, write notes in the margins and by all means, ask questions. When do you want the paint system, supplier, and installer identified?" he asked.

"A year from now," said Kanisha.

"Six months?" asked Cheryl.

"Don," Frank asked, "when do you assemble your capital plan for next year?"

"August," came the reply.

"Will you need to get capital approval for this project?" Frank asked.

"Absolutely," said Don.

"Finally," Frank concluded, "when do you want the new system installed?"

"Next year," replied Don, while looking at Cheryl.

"Oh, sugar," came her reply.

"Guess when you'll need this study completed by?" queried Frank.

"August," said Manny, "before we submit the capital plan."

"So," Frank responded, "you figure you'll be able to accept the team's recommendations as soon as they give them to you? You won't need any time to send them back to the drawing board?"

Manny looked unsure.

"Trick question," announced Frank. "You will be ready because you'll be getting monthly updates on the status of the study. So, let me ask again, when do you want the team to have the final recommendation to you?"

"August?" asked Manny.

"August!" announced Frank with conviction. "So far so good?" he asked, looking at Cheryl.

"I'm OK if they are," she announced. "What do you think? Who has a problem so far?" she asked. "Raise your hand."

No one raised their hand. She looked back at Frank. "No problems," she announced.

"Moving on then," Frank continued, "how about these?" pointing to the box below.

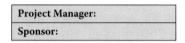

"Project manager will be Samantha Brown," said Don after consulting his notes.

"Does everyone agree?" Frank asked.

"Show of hands for agreement," announced Cheryl. It was unanimous.

"Very good," announced Frank. "Who's the sponsor?" he asked. This time he was met with silence. "Let's come at this another way. How high in the organization should the sponsor be?"

"High enough to understand the paint process and its impact on the rest of manufacturing," offered Kanisha.

"So, at what level do you think that occurs?" Frank asked. "Are we looking for a supervisor, a manager, or a director?"

"Director, for sure," responded Kanisha.

"OK. Then what director do you think is best suited to be the sponsor?" Kanisha looked at Cheryl. "Thanks," said Cheryl sullenly.

Kanisha put it to a vote and Cheryl received unanimous approval. Frank added Cheryl's name to the sponsor box.

"Next," Frank continued, "We're going to record the background of the problem you'll be examining. Simply put, you'll list what the conditions or problems are that you're examining and why. So, what are we asking the team to examine, and why?"

Background
What problem(s) are you examining and why?

The team spent almost 15 minutes working through a definition of the problems that led to the need for a new paint system. As they worked, Manny kept notes. When they finished, Frank asked Manny to type the background statement into the form on his laptop.

| Background | The painting system at Friedman Charleston is over 20 years old. Its setup and throughput times are slow, and pace the entire factory. Its cleaning system is much less effective than newer ones and the painting system itself contributes to a high number of defects in painted parts. Maintenance on the current system is costly, parts are hard to get and the system is down approximately 15% of the time needed. |

"See how easy this is?" Frank asked. He was met by groans.

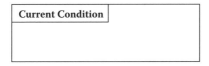

"Next are the current conditions. You can handle this a number of ways. Some people put a flowchart or future state VSM in this space, showing the starbursts of areas needing work. Others might list parts of the system that are not ideal. Still others list the current state of KPIs, their present values and goals. There's no wrong answer, but I find that getting the team to concentrate on KPIs is a good way to keep them focused on what you want them to improve. Now, what do you want to list as your current conditions?" asked Frank.

Having already been through the brainstorming of the background, the current conditions went much faster. There was some talk about cutting up a reduced copy of the VSM and pasting it in the space, but Manny pointed out that they were only interested in the paint area and it only reflected one block of the plantwide VSM.

"Really," Cheryl stated, "all we really care about are the key characteristics of the system. Those are," she paused, flipping through her notes, "defects, cycle time, changeover time, downtime, and work in process. Let's just use those. Besides," she continued, "it will make it easier for them to track and us to follow. Not that I'm lazy, you understand," she admitted with a smirk. That drew smiles all around, but the team agreed with her logic and posted the current state key characteristics.

Current Condition	• Changeover Time = 4.5 hrs
	• Cycle time = 3 hrs and 20 min
	• Downtime = 15% of available time
	• Defects = 10%
	• WIP = 1,200

When the current conditions block was filled out, Frank added, "By giving the project manager these data, you've established a baseline for them. When you give them their goals, they will understand the magnitude of the challenge you've issued them. Filling the current condition box with baseline data works in this case, but it may not always. Under other conditions, you may want teams to diagram the "as is" problem, or paste their value stream map into the box, or explain what about the current condition is problematic. In short, unless you guide them otherwise, the team will have a lot of flexibility in how they fill out this form."

Shifting gears, Frank said, "Let me remind you of a conversation we had months ago. What do we call the kind of event that tackles changes to whole systems?"

Chris, Don's IT director, rifled through the pages of a three-ring binder he'd brought with him. Soon the rustling stopped and he announced, "A Kaikaku event."

"Very good," Frank praised him. "While you're there, Chris, remind us about some of the distinctions between Kaizen and Kaikaku events."

Reading from his notes, Chris said, "Kaizen events are short, usually five days or less. They are very focused and have a narrow scope. They include all the stakeholders of a single process, including representatives from the supplier and customer operations. Kaikaku events," he continued, "tackle whole systems, they require participants from all the disciplines that are affected by the system, and they take a long time. Kaikaku events typically result in a strategic advantage over one's competitors."

"Excellent," Frank praised Chris again. "I can see you take good notes. Which type of events do you think we're looking at here?"

Chris thought for a second, "Kaikaku?"

"Correct," replied Frank. "The thing that makes these tactical objectives Kaikaku-worthy is that they span multiple disciplines and will have life cycles of multiple months. By the way, can you see how this year's tactical events will give you a strategic advantage? Something to keep in mind is that many of these tactics will, most likely, break down into subordinate events. These subevents may not be Kaizen or Kaikaku events, but you will need to create project teams that will tackle issues such as the effects the tactic will have on quality, plant engineering, purchasing, EHS (environmental, health and safety), and so on. Remember that conversation? Those project teams will then report back to the tactics team and keep them apprised of what they're learning."

"Just like the top-down metrics we learned earlier, this is a system of accountability. Each team reports its progress upward, each adding to the overall company goal, and the efforts of each coming into clear focus at your monthly briefing." Frank flipped the goal/target box on the screen. "What do you think goes in this block?" he asked.

"Our tactical goals," Don stated.

"Very good," Frank said. "So, what are they?"

This time Don flipped through his notes. "Defects ≤1.5%, CT ≤15 minutes, changeover time ≤5 minutes, downtime = 0%."

Goal/Target	Target	Actual
	• Changeover Time ≤ 5 minutes	• 4.5 hours
	• Cycle time ≤ 15 minutes	• 3.3 hours
	• Downtime = 0%	• Downtime = 15%
	• Defects ≤ 1.5%	• Defects = 10%

"Very good," Frank praised the plant manager. "Just a reminder," he continued. "I've introduced you to what's called a seven-box A3. There are other versions that have nine and more boxes. I chose seven for brevity. After you've worked with the seven-box version for a while, you can always change, but I'd encourage you to do two things: first, I encourage you to use the seven-box for at least the first year. Second, I'd strongly suggest that whatever A3 format you choose, you convert your entire organization to it. I do know this," he continued with a smile, "corporate and the other factories all use the seven-box format."

"From this point on you're going to be pretty much on your own, as no data currently exist to fill in the rest of the boxes. What I will do for you is to give you some examples of what might go in each box. Remember, the more structure you can give your teams in the use of this form, the more consistent the reporting will look; however, the more structured you make the reporting, the less flexibility you give your teams to think outside the box and to develop new and unique ways of displaying the results of their work. OK, what do you think will go in the analysis box?"

When no one volunteered an answer, Frank said, "This is where the team will list the suspected root causes of the current condition. What techniques can the team use to get to the suspected root cause of a problem?"

"I've got this one," said Jennifer. "The first is to ask why five times. The other is to use a fishbone diagram."

"Very good," praised Frank. "So what might you expect to see in the analysis box?"

"I'm going out on a limb here," offered Cheryl a bit tongue-in-cheek, "but my guess would be their 5 Whys or fishbone diagram."

Frank nodded. "Now, for the recommendations box. We're going to list any one of a number of things here. For instance, you could list the target condition you or the team want to achieve."

```
┌─────────────────────────────┐
│ Recommendations │           │
│                             │
│                             │
│                             │
│                             │
└─────────────────────────────┘
```

"In the current case, you've already listed those in the goal/target box, so you might have the team put their future state value stream map here. The starbursts will indicate the ideal state or process values. Of course, you could do the obvious and ask the team to give you their recommendations to improve the current condition. These can't be wild guesses or hypothetical recommendations, though, because in the next box, they're going to have to articulate their specific plan to achieve each of their recommendations."

```
┌─────────────────────────────┐
│ Plan │                      │
│                             │
│                             │
│                             │
│                             │
└─────────────────────────────┘
```

"How about the plan box?" Frank asked. "What goes in there?"

That drew a chuckle from the team.

"Actually," Frank carried on, "this has to be very specific. In this box the team needs to identify what they will do, who will do it, when it will be completed by and, to the extent there is a location at which it will be performed, they'll need to note where. They'll also be expected to give the cost of each action. So, the plan box might look like this." He drew the figure on the board.

Plan				
What	**Who**	**Where**	**When**	**Notes**
Proposed countermeasures	Who will lead implementation?		Completion due by?	List changes, discoveries, reasons for missing completion date (always with a recovery plan), cost to implement, etc

Frank noted that the plan box would typically have multiple horizontal lines dividing each column of data into as many boxes as needed to cover all countermeasures and their associated data. "And last," Frank continued, "we have the follow-up box."

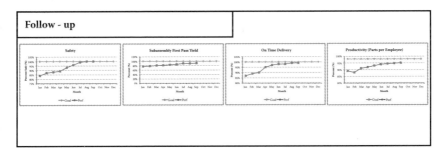

"Here they'll account for their achievements against plan and what further action they'll need to take. In this box, I recommend that teams post run charts depicting their performance against their original goals. That will mean that your run charts will need to shrink and/or your box size will need to grow." He projected an example. Then he stopped and just let silence settle on the room. His face was turned to the floor, indicating that he was thinking. Finally, he looked up, still silent. "That's a lot," he said at length. "How are you feeling about all this?" There was silence, and Frank let it build. He had no intention of moving on until he got feedback.

Finally Don spoke. "You're right. It's a lot. On one hand I feel intimidated, wondering about things like what if we mess up? On the other hand

I'm thinking this could finally give us the focus we need to become the best in our market and how could messing up be any worse than doing nothing? After all, to this point we've totally ignored our future, as if it didn't exist. The fact that we will finally be addressing it excites me."

Frank nodded his head, but said nothing.

Kanisha raised her hand and Frank pointed to her.

"I've been with companies that had strategic plans. They were a waste. I can see how this could be different, but know that success is all about follow-up. Have we talked about that?" she concluded.

"Glad you asked," Frank replied with renewed gusto. "That's where the rubber meets the road, isn't it, follow-up? I'm not going to answer your question, though. It's not mine to answer. Let me turn it around and ask you, who will follow up?"

Don didn't wait for the others to answer. "That's mine," he said, "but I'd be grateful for your insights on how best to do it."

"Thought you'd never ask," Frank smiled at him. "By the way," he continued, "the rest of you should stop breathing a sigh of relief and recognize that Don just did exactly what he needed to, he stepped up. He has taken ownership of the burden that's his. That is truly why he gets paid the big bucks." There were chuckles around the room, then Frank continued. "As for next steps, here's what I recommend."

"**Step 1.** As a group, meet with the project managers. Give a high level explanation of the X-matrix and for which tactic each will be responsible. Before you do that, you need to decide whether you'll allow those you've chosen to decline the job. If you will allow them to do so, you will want to tell the project manager group that you'll meet with each of them individually to get their answer, but that this is a growth opportunity. Remember the people from other Friedman plants who were here during the transition period? Guess how they got chosen?"

"Let me digress on that point. Unless you really have plans for promoting the project managers in the future, I would not tie any career outcome to the successful completion of this assignment. As I just said, you are giving these employees an opportunity to grow and showcase their talent. That can only help them. If you want to give them a one-time bonus for the successful completion of the project, that's up to you, but they should not get an ongoing increase in their pay. If they will only accept the job if you give them more money, they're likely the wrong people."

"OK, back to Step 1. During this first meeting, you'll want to disclose that the A3 will be the tool used to report and that they will receive specific training on it, as well as ongoing coaching. You'll want to make it clear that they will not be allowed to fail at the task of reporting. Questions?"

No one had any.

"**Step 2.** A3 training. You'll want to conduct A3 training for all the project managers and their sponsors. Those disciplines may have different questions. You'll want to tell them that you'll hold a monthly meeting to review progress against their project, and that they'll brief from their A3.

"**Step 3.** The monthly meeting will be a formal meeting that all of you and all of the project managers and all of their sponsors will attend. One by one, they will be asked to brief their A3 and answer your questions."

"I'd recommend that you create a wall or walls somewhere in a conference room where all A3s, and any associated documents can be posted. This will need to be a permanent site and the items posted there will remain there until they are updated. I'd even recommend that you go so far as to outline the area with tape, and place the tactic name at its top. Keeping 5S in mind, you'll want this area and all postings to be neat, clean, and follow a specific format."

"If Don is not going to emcee the meeting, you'll also want to appoint one person from this group who will own the monthly briefing. You can make an argument for doing it either way. I'd recommend that you start with Don running the meeting, but it's not an issue of manhood for me. The meeting itself will be a stand-up meeting. That will make it clear that you aren't looking for long briefings or lengthly discussions. Whoever runs the meeting needs to be quick to deflect sidebar conversations offline. This meeting is a rapid-fire update. After the meeting, each program manager will receive individual coaching by his sponsor."

"**Step 4.** Conduct monthly meetings on schedule and stick to the time allocated. I'd encourage you to put these meetings on your calendars and schedule everything else around them. Keep in mind, you'll be discussing the future of your business. There won't be many things more important than this. Still, get them on peoples' calendars as far out as possible so they can schedule vacations and other engagements around them."

"**Step 5.** Recovery plans. The future success of your company will be tied to the timely accomplishment of these tactics. Sometimes, the success of multiple tactics will be linked, so that if the project manager on one misses a due date, the success of another will be adversely affected. You'll need to make it clear that every effort needs to be made to meet milestones on time."

"If a project manager believes he will miss a milestone, he needs to disclose that fact to his sponsor as soon as he discovers it. That concern will need to be elevated immediately so that remedial action can be discussed and swiftly implemented, if indicated. If a milestone is actually missed, the project manager needs to create a recovery plan that gets the project back on schedule. These can't be based on a wish and a hope. They need to be based on fact and hard work."

"**Step 6.** Things rarely go as planned, so this group will want to hold quarterly meetings to assess drift and readjust your own plans."

Frank stopped and just looked at his audience. "Questions?" he asked.

There were none.

"All right," he said, looking at Don. "You know how to reach me." With that he started packing up and the team filed out.

18

The Long March

In the months that followed, performance at the Charleston plant tightened considerably. All of their top-level KPIs continued to improve. The study for the paint system had been completed, a design chosen, and an installer identified.

The team studying the relocation of the paint system had concluded that it would only cost a small percentage more to relocate it, but that the benefits would far outweigh the costs. They created a 3D model of the new layout and demonstrated that it was reasonable to expect a day to a day and a half reduction in total product throughput time.

The capital plan was submitted to include both the new paint system and the relocation. Because the project teams had been so thorough in their documentation of the advantages, the plan was approved and work began within three weeks.

The new location chosen for the paint system was in the old storage area (more a junk pile) at the end of the mechanical assembly line. Because the design team had avoided using any manufacturing space, the new paint system went in while the old one was still running.

After the construction crew went home each day, all employees were invited to observe the progress. There had been an easel and flip chart placed in the observation area where employees were invited to make comments and suggest modifications. The comment board soon turned into a cheerleading station. There were praises written for the contractors, for the design team, and for the paint group in general. But there were also suggestions for improvement that frequently got incorporated. There was a general buzz of excitement in the plant.

Although a couple of painters were on the project team, all the painters were particularly interested in the new system. At Friedman's request, the company installing the system set up a small training station where, every night after their shift ended, Friedman Electronics paid the paint team overtime to go through an hour and a half of training.

The painters were trained in the new safety gear, the hazards of the new paint, the use and cleaning of the gun, the operation of the new system, and troubleshooting tricks. By the time the system was officially opened, the painters were already up to speed. The last week, every painter actually used the training system to paint parts.

Friedman maintenance personnel were also trained in the care and maintenance of the new system. A total productive maintenance (TPM) Kaizen was held to establish the TPM schedule for each component in the system. TPM locations were identified, numbered, and photos taken of each. The photos were then added to a sheet called a dashboard. Adjacent to each photo were instructions on what action took place at that location and how often. When finished, these dashboards were laminated and placed in the painter's daily action log. Each painter needed to complete the daily TPM before they began painting that shift.

A separate 5S Kaizen was conducted to purge the maintenance crib of all the spare parts associated with the old paint system. The same team cleaned and labeled the shelves before installing the spare parts for the new system. The old system had become so dilapidated that its spare parts occupied an entire row of shelving. When its spare parts were removed, the team was able to eliminate the entire row freeing up almost 200 square feet of prime manufacturing space. Although the system was old, Francine found buyers for all the spare parts. It seems that other companies were still trying to limp along using outmoded technology.

Of course, the real point of the installation was to come up with a better painting system. In that regard, the new system exceeded expectations. It included a totally novel paint delivery process. The painter stood in the booth with the gun. Compressed air was delivered to the gun, but there was no bulk paint tank or long, heavy delivery hoses. Paint was stored on a metal tree equipped with hard plastic bottles. The bottles were transparent so that the color of the powdered paint inside could be seen through them. The bottles were fitted with special, quick-release caps that allowed them to be removed from the gun while there was still paint in them. The spring-loaded cap prevented any of the powder from escaping, but snapped in and out of the gun for rapid loading and unloading.

The gun was designed so that the painter could finish painting one color, pop that bottle out, place it back on the tree, grab the next color, load it, and be back painting a completely different color in about 15 seconds. Setup time was now negligible, and the lot size had been reduced to one.

The long overhead conveyor and ovens had been replaced by a single-story infrared oven. The length of time between hanging the part and removing it was now less than 20 minutes. The oven was arranged in an oval. It made a 360-degree turn, returning to within 20 feet from where it had been loaded. That placed the offload station 20 feet from the load station, allowing for the flexing of personnel from one location to the other as conditions required.

Because the paint system had been installed adjacent to the rest of the manufacturing line, products came out of mechanical assembly, were hung, painted, removed, and placed right into final electronic assembly. It was as if paint were no longer an obstacle to the rest of manufacturing.

Defects were another thing. Despite the training, for the first two weeks, painters had difficulty making the change from wet to dry processes. The mechanics were easy, and the painters had adapted quickly to the rapid change of colors and difference in gun styles. What had proven difficult had been mastering an understanding of how the new paint dried. Rather than fight through the problem, Don had suggested that Cheryl bring the training crew back in. It was an added expense, but within a week, defects had dropped to less than 1%.

At the next monthly meeting, the two paint project teams were applauded as a group. It was evident from their blushing that their peers' appreciation meant a great deal to the team members.

Although the paint system selection team had never agreed to measure it, volatile organic compounds from the new system were a fraction of the old system's. Kanisha and Manny put their heads together and found a way to sell the difference to another company for a year, knowing that they would likely need to use them when the new plastic cabinets went into production.

On that token, market studies had discovered that consumers wanted the strength and rigidity of sheet molding compound (fiber reinforced plastic). That eliminated technologies such as extrusion, rotational molding, blow molding, and casting. The same market studies indicated that consumers wanted cabinets as deep as 25 inches and as tall as 84 inches. Most, however, wanted the height and width of the cabinets to be achieved through the assembly of a series of smaller cabinet "tubs."

Tests run at equipment manufacturers proved that injection molding could not handle the deep draw required to make the tubs. Compression molding did much better, but every configuration required a different die and they could run over $1 million each.

Finally the team accepted one of Kanisha's recommendations that they make each individual wall of the tub independently, then assemble the components. The team sought the advice of fastener companies, but actually got inspiration from one of the manufacturers who was running their tests.

The manufacturer showed them how they could mold elements into the panels so that, when assembled, mating panels could be snapped together. Likewise, the hinge could be molded right into the door and side panel. Now, instead of inserting multiple fasteners to hold the cabinet together, the assemblers only needed to snap the panels together and install a gasket on the door. Furthermore, the manner in which the components were molded made them nearly watertight when snapped together. More good news: by making only flat panels, the size of the press required was dramatically reduced, which dropped the cost of the project by almost 75%.

The parts were now small enough that opposing sides of the box could be pressed in the same mold. That meant that all six sides of the boxes could be molded simultaneously in three separate presses. Although not the perfect location, the area freed up by the paint booth and storage space reductions would easily accommodate the new presses.

If arranged properly, the output end of the final press could be located almost adjacent to the inlet of the paint system, where they could be painted before subassembly. That meant that both plastic and metal boxes could have their electronic innards assembled on the same line.

All teams had representatives from Francine's materials' organization. As new materials or equipment were discussed, these representatives found potential suppliers, had them send specifications, samples, and even arranged plant visits when appropriate. By conducting a weekly meeting with her people who were on the teams, Francine kept her finger to the teams' pulse.

Cheryl also held weekly meetings with her team members. By doing so, she not only demonstrated her support, but gave them practice using the A3. Always the politician, Cheryl realized this was a good way for her to know what was going on and ensuring her people looked good at the monthly stand-up meeting.

19

The End of the Beginning

By the end of the second year following the Hoshin, the new molding system was in place and Charleston had begun to make plastic cabinets. They'd had their share of startup problems. They had begun with the fact that no one in Charleston made sheet molding compound (SMC). The compounder they intended to use was the one near the press manufacturer, where they'd run their trials. Unfortunately, that had been in Ohio, and by the time the compound reached them it had been dry.

Dry compound, they'd been warned, resulted in cracks in the parts and, sure enough, it had. Of the first compound batch they'd ordered, less than 50% could be used. That had been a costly problem. Don and his team had considered making their own SMC, but the costs of entering that business were huge: massive resin tanks, large silos of dry compounds, pumps and piping to convey the powders and resins, creels of fiberglass, mixing vats, and, of course, the compounding machine itself. It wasn't just the equipment cost. The other contributing factor was that compounding wasn't part of Friedman's forte and they didn't want it to become so, at least not now.

Francine went back to the supplier. Working together, they'd devised a special packaging system that was airtight and recyclable. Plus, they now placed each roll of SMC in a plastic bag and sealed it. The trick had worked. Soon Charleston became the first Friedman plant making plastic cabinets and try as they might, they couldn't satisfy the demand.

Jennifer was excited that plastic cabinets were backordered three months out. Don's boss, Jim, had been in to see the operation, and there had been discussions about adding the capability to make plastic cabinets at one of Friedman's California plants.

Soon Friedman employees were streaming to Charleston to see their operation. During one of the visits by a new plant manager, Don had a déja vu moment. In that moment he'd recalled himself in Jorge's plant.

Although pleasant, that memory seemed like a thousand years ago. Reflecting back, Don was humbled to realize what an arrogant and self-righteous so-and-so he'd been. He recalled how kind and selfless Jorge had been by contrast. He'd realized, even then, that he'd wanted to be that kind of a manager. Of course, he'd never be Jorge, but since his visit to Oakland, he'd become a much better version of himself. He realized that he owed that to Jim, who had believed in him even at his worst.

Why Jim had kept him, when he'd clearly been an arrogant stuffed shirt, continued to mystify him. He realized, once again, that he owed Jim a great deal. In subsequent visits to Charleston, Don and Honey had entertained Jim, and not at the country club. Honey had actually cooked for him. She'd been a wreck; you'd have thought she was cooking for the Pope, but Jim had been typically gracious and put her right at ease.

Before supper, Jim had insisted that Don take him for a tour of the property. Drinks in hand, Don had been proud to show Jim the nursery and introduced him to his nurseryman. On the way back in, they'd gathered vegetables to bring in to Honey. Jim had been highly complimentary of what Don had done to the property and Honey to the house.

The meal and conversation had been convivial. When he left that night, Jim had praised Honey for her cooking and decorating. He'd elbowed Don and said he couldn't understand why they'd ever have entertained at the country club. The compliment had pleased Honey and she blushed deeply.

Now in his office, Don withdrew a sheet of stationery from his desk and started a note to Jim. He could have sent an e-mail, but knew a handwritten note would do a far better job of conveying the gratitude he felt. He thanked Jim for believing in him. He thanked Jim for helping him pick such a great new staff, and for coaching him in the new way to lead.

Don recalled returning from Jorge's plant to find that Jim had fired his staff and was already interviewing new people. He'd been so angry he'd challenged Jim's authority. Jim had almost fired him, too, but for reasons Don would never understand, Jim hadn't.

Looking back, Don realized that he could never have gotten here with his old staff and that Jim had been right, Don had surrounded himself with people who made him feel important. He'd run a Friedman plant as if it were a plantation and birthright. What an ass he'd been. He was in awe of Jim and his patience. He told him so in the note.

By the end of his note, Don felt newly humbled. He wondered what he could do for Jim that would somehow convey the gratitude he felt. Then he recalled something that Jim had once said to him: "Don, if you really like what we've done, do your best to emulate the style of leadership you've learned. Remember, I'm expecting you to build a deep bench here in Charleston."

Don smiled as he signed his note.

"Building Bench in Charleston."

Don

Index

K

Kaikaku
  events span multiple areas, 138–139
  process, 104–105
  system-wide, 106
Kaizen events, 12; see also Lean events
  to address deficiencies, 20
  authority to set in motion, 23
  continuous improvement (CI) staff
    role, 42
  implementation of, 105–106
Kaizen literal translation, 104
Kanri, definition, 51
Key characteristic monitoring, 62
Key performance indicator (KPI) boards
  chart example, 84
  for each manager, 83
  metrics, 75
  printing of, 85
  for problem analysis, 52–53
  reveal organizational problems, 103
  use of, 51–52, 57
Key performance indicators (KPI)
  derivative, 125
  points to improvement opportunities,
    121–122

L

Lagging indicator, definition, 125
Leadership
  distributive, 72–73
  guidance of subordinates, 1–2
  styles, 3–4, 27, 29–30
Leadership behavior to maintain Lean, 12
Leadership decision levels, 23–24
Leadership vs. management, 27
Leading indicator, definition, 125
Leading vs. managing, viii
Lean
  respect for people, 86
  as way of leading, vii
Lean assets, definition, 40–41
Lean council role, 42
Lean education and training, 41–42
Lean events; see also Kaikaku; Kaizen
    events
  function of, 42

wrap-up process, 43
Lean initiative
  maintaining effort, 12
  at Oakland plant, 33–35
  power dispersal, 37
Lean leaders necessary for Lean, vii
Learning to See, 16
Liker, Jeffrey, 41
Long-term philosophy rather than short-
    term financial goals, 41
Lowest cost of ownership, 57

M

Maintenance efforts via total productive
    maintenance (TPM), 146
Management bench; see also "Building
    bench"
  building, viii, 13
  failure to build, 25–27
Management knowledge of subordinates,
    91–92
Management responsibility for
    breakthrough success, 142–144
Management style
  autocratic, 3–4
  formal vs. informal, 29–30
Management vs. leadership, 27
Managing vs. leading, viii
Manufacturing metrics, 76
Marketing metrics, 76–78
Materials metrics, 76
Meeting effectiveness, 71
Meeting purpose, 73
Metrics
  accountability for, 50
  cascading, 103
  expectations expressed in, 128–129
  key performance indicator (KPI), 75
  manufacturing, 76
  marketing, 76–78
  materials, 76
  roll down to lower levels, 75–76
  rolled down to floor level, 51
Missed milestone recovery plans, 143–144
Mistake-proofing techniques, 104
Motivation
  avenues, 91
  background for understanding, 96